PENGUIN BOOKS

SPACE

Tim Peake is a British astronaut. Tim completed his
186-day Principia mission to the International Space
Station with the European Space Agency when he
landed back on Earth on 18 June 2016.

He is also a test pilot and served in the British
Army Air Corps. Tim is a fellow of a number of UK
science, aviation and space-based organisations and a
STEM ambassador. He is married with two sons.

Tim's bestselling books include *Hello, Is This Planet Earth?*,
Ask an Astronaut, *The Astronaut Selection Test Book* and *Limitless*.
His books for children include *Swarm Rising*, *Swarm Enemy*
and *The Cosmic Diary of our Incredible Universe*.

PENGUIN BOOKS

UK | USA | Canada | Ireland | Australia
India | New Zealand | South Africa

Penguin Books is part of the Penguin Random House group of companies
whose addresses can be found at global.penguinrandomhouse.com

Penguin
Random House
UK

First published by Century in 2023
Published in Penguin Books 2024
001

Typeset by in 10.56/13.64 pt Goudy Old Style Std by Jouve (UK), Milton Keynes
Printed and bound in Great Britain by Clays Ltd, Elcograf S.p.A.

The authorised representative in the EEA is Penguin Random House Ireland,
Morrison Chambers, 32 Nassau Street, Dublin D02 YH68

A CIP catalogue record for this book is available from the British Library

ISBN: 978–1–804–94626–8

www.greenpenguin.co.uk

SPACE

The Human Story

TIM
PEAKE

PENGUIN BOOKS

'*If I have seen further [than others], it is by standing on the shoulders of giants.*'

Sir Isaac Newton, 1675

To all the giants, who keep searching for answers.

CONTENTS

INTRODUCTION

On 14 December 1972, at the end of a seven-and-a-quarter-hour shift, Eugene Cernan lifted his foot onto the bottom rung of the Lunar Module's short aluminium ladder and began to haul himself up. He had thought hard about what he would say at this moment, knowing what it signified.

'We leave as we came,' he began, his voice crackly and breathy over the intercom, 'and God willing, as we shall return, with peace and hope for all mankind. As I take these last steps from the surface for some time to come, I'd just like to record that America's challenge of today has forged man's destiny of tomorrow. God speed the crew of Apollo 17.'

Cernan had been camped on the Moon for nearly three days at this point, and he would be there for a little while yet. Once inside the module, with the hatch secured and their dust-covered spacesuits and oxygen packs removed and stowed, he and his colleague, Harrison 'Jack' Schmitt, still had a long and complicated list of preparations and pre-flight checks to complete, not to mention a scheduled eight-hour sleep to fit in – their last spell of rest in the tiny metal container which had been their shelter a quarter of a million miles from home.

Eventually, though, it was time to go. On the ground at Mission

Control in Houston, amid the banks of screens and the plumes of cigarette smoke, the usual atmosphere of taut concentration was mixed on this occasion with something more poignant. To the disappointment, frustration and even outright dismay of nearly everyone at NASA, there would be no more government funding for the Apollo lunar programme beyond this mission. After a hectic three-and-a-half-year period in which America had landed six pairs of astronauts on the surface of the Moon, the plug had been pulled. When Apollo 17's lozenge-shaped capsule eventually descended into the Pacific Ocean under its three red and white striped parachutes, it would also be bringing down the curtain on human lunar exploration. When, if ever, that curtain would go up again, nobody knew. With calculated irony, Mission Control chose to rouse Cernan and Harrison from their slumbers by piping into the module the sound of the Carpenters singing 'We've Only Just Begun'.

'Now,' Cernan said, when the checks were done, 'let's get off.'

He entered the Proceed code into the on-board computer, the ascent engine ignited, and, in a flash of light, the module lifted off the surface and rose into lunar orbit. And with that, the astronaut who would be known for the rest of his life as 'the last man on the Moon' began to make his way back to Earth, and the most astonishing chapter to date in the story of human exploration came to an end.

Now come forward fifty years, to 11 December 2022. Half a century to the day since Cernan and Schmitt landed on the Moon (and a little under six years since Cernan's death at the age of eighty-two), another tiny cone-shaped spacecraft floats down from the sky under three striped parachutes and drops into the Pacific Ocean, this time just west of Baja California, where the waiting USS *Portland* moves in to begin the recovery process.

This is the conclusion of the Artemis I mission. For those whose

Greek mythology is a little rusty, Artemis was the twin sister of Apollo, and that capsule now bobbing in the water is just back from a 1.3 million-mile voyage in deep space, including six days on a distant retrograde orbit of the Moon and two lunar fly-bys, passing just 80 miles above the Moon's surface. Amid temperatures rising to 2,700°C, it has just plummeted back into the Earth's atmosphere at a speed in excess of 24,000mph, before those parachutes opened to break the capsule's fall and eventually bring it floating down to the water at a more sedate 20mph.

In the Commander's seat: Captain Moonikin Campos. On either side of him: Helga and Zohar. The Captain is a male-bodied mannequin named after Arturo Campos, the NASA engineer who played a major role in the rescue from disaster of Apollo 13.* Helga and Zohar are female-bodied model torsos, named by the German and Israeli space agencies respectively. Also on board: a small plastic Snoopy and, flying the flag for the European Space Agency, who are also involved in this multi-national project, a model of Shaun the Sheep.

No humans, then. But that's for next time. The follow-up mission, Artemis II, scheduled at the time of writing to launch in late 2024, won't land on the Moon either, but it will attempt a final trial journey with four astronauts on board, much as in May 1969 Apollo 10 (with, incidentally, Eugene Cernan in its crew) closed to within 9 miles of the Moon's surface for one last recce before Neil Armstrong's Apollo 11 went all the way and

---

* Mannequin, Moonikin . . . you'll see what they did there. Mannequins have a distinguished and perhaps under-celebrated role in the history of spaceflight. Later in this book we'll meet Ivan Ivanovich, who nervelessly tested Russia's Vostok capsule before Yuri Gagarin flew in it. Ivanovich was so eerily lifelike that he had to have the word 'MAKET' – 'DUMMY' – inserted under his visor so that no one would confuse him with the real thing.

landed. And then, all being well, the path will be laid for Artemis III to take humans to the surface, possibly in this decade.

At which point, Eugene Cernan will be posthumously relieved of the title he never wanted to hold by the next man on the Moon – who, incidentally, will most likely be a woman.

I was only eight months old when Cernan left the Moon in 1972, and if the moment made little impression on me at the time that's because I was rather more concentrated on a project closer to home, namely how to get upright on my legs. Later in childhood, though, I would feel the lasting resonance of that whole set of Apollo missions, as so many of us did and continue to do. I was by no means a space nut early in my life: I was more fascinated by things that flew closer to the ground, like helicopters. Nevertheless, like so many kids my age, I clicked together my LEGO Saturn V rocket and launch tower and thought with awe about Neil Armstrong and his 'one giant leap for mankind'. Even after the Apollo programme was shut down, the crackle and beep of astronauts calling down to Earth from space were as much a part of the soundtrack of a seventies childhood as ABBA songs and Scooby Doo's snigger. Time and again you would be confronted by the scale of the Apollo project's audacity and experience its capacity to pull you up short and cause you to catch your breath, a capacity it retains even now, half a century later.

Years further on, while training as an ESA astronaut, I would be afforded the privilege of working at the Johnson Space Center in Houston, where, inevitably, spaceflight's pioneering past was never far away. I would get to tour the preserved Apollo-era Mission Control room – and discover that it still smells of cigarette smoke. And I would be taken on a familiarisation visit to Cape Canaveral, and be shown round NASA's Vehicle Assembly building, which rises like some kind of monster out of the Florida scrub – 525 feet tall and spread over 8 acres, the largest single-storey construction in the world, a building so vast that it routinely defeats attempts to photograph it. At one point on my visit I was taken up in the lift to the top of an Ares rocket, which was then being prepared. The rocket was just over 300 feet tall and reaching its summit took a while. Looking up, I realised that was only just over halfway to the ceiling.*

It was impossible to spend time in those places and not be made powerfully aware of the Apollo legacy. It was all around you. Similarly, I would come to feel the presence, too, of the Russian pioneers, who during the 'Space Race' of the 1950s and 1960s beat the Americans to so many spaceflight landmarks – first functioning satellite, first man in space, first woman in space, first spacewalk, first three-person crew in space – if not, ultimately, to the Moon. My journey to the International Space Station was on a Russian Soyuz rocket launched from Baikonur in Kazakhstan, where Yuri Gagarin's history-making journey began, and every astronaut who flies from there pays a series of tributes to Gagarin before they depart. I laid red carnations on Cosmonauts Avenue outside the wall of the Kremlin in Moscow, where Gagarin's ashes reside. I paused in Star City to sign the visitors book in Gagarin's old office, preserved as he left it. I got my hair cut two days before the launch, as Gagarin did, and signed my name on the door of my room at the Cosmonauts Hotel, just like Gagarin. And on the shuttle bus to the launch pad, and by now fully suited, I nevertheless got out to unbutton myself and take part in the ceremonial bladder-emptying by the bus's rear right wheel – because, yes, that's what Gagarin did, caught short on that morning in April

* That Ares 1-X rocket was a prototype, and was launched a couple of months after I saw it, in 2009, on a sub-orbital test. It was part of the President Bush-era Constellation programme which was intended to develop two rockets – an Ares 1 to carry the Orion spacecraft and an Ares V to carry cargo. But the programme was cancelled in 2010 so in the end I saw the first and last Ares 1.

1961. At a ceremonial breakfast on the morning of my departure from Star City for Kazakhstan in 2015, I had found my hand being energetically shaken by Alexei Leonov, then eighty-one years old, Gagarin's colleague, the first man to walk in space, and who very nearly lost his life while doing so. (We'll come to that story.)*

Later, as a flown astronaut, in a development that my childhood self would have found entirely implausible, I got to meet and spend time with some of those Apollo astronauts – with Buzz Aldrin, Charlie Duke, Al Worden, Walt Cunningham, Harrison Schmitt and Rusty Schweickart. Indeed, once, at a space convention in Zurich, I knew the completely withering experience of being obliged to do a public speech in front of a selection of Moonwalkers.

Tough crowd.

What with one thing and another, the Apollo programme has resonated loudly throughout my life. But I never had that experience which people who were old enough at the time speak of: the experience of looking up at the Moon, during a mission, across that vast distance, and knowing that, at that actual moment, humans were up there, on that utterly separate surface. That, people will tell you, brings the full magnitude of it home. The curtain is about to go up again, though, and soon I and a whole new generation will get to gaze up into the sky and finally know what that sensation feels like.

'Those steps up that ladder, they were tough to make,' Eugene Cernan later said. He had taken a moment to look back over his shoulder where he could see the Moon's mountains and, beyond them, the Earth. 'I didn't want to go up. I wanted to stay a while.'

---

* Alexei Leonov died in 2019, aged eighty-five. He was the longest surviving of Russia's original draft of Voskhod cosmonauts.

Cernan and Schmitt spent seventy-five hours on the Moon, the longest of all the Apollo stays, and were out on the surface for over twenty-two hours, in three separate excursions. As with the two Apollo landings before them, he and Schmitt had access to the Lunar Rover Vehicle, NASA's Moon-ready golf buggy. At one point they were 4.7 miles away from the safety of the module, at which time they were arguably as remote and as exposed as any human explorers had ever been.* They discovered orange-coloured pieces in the soil which proved to be beads of volcanic glass, sprayed as lava from the Moon's core in a fire fountain which probably happened 3.5 billion years ago, and they brought back samples that directly altered our understanding of the Moon's formation. Cernan did not, as frequently related afterwards, scratch his daughter's initials into a rock which he and Harrison had examined in the Taurus-Littrow valley – a huge split slab, the size of a house, which they named Tracy's Rock. But he often wished afterwards that he had done so, and said as much to his colleague Al Bean, who flew on Apollo 12 and also walked on the Moon. And Bean, who took up art on his return to Earth, duly painted a picture which showed the valley in carefully accurate detail, but with additional graffiti, 'TRACY', putting the record straight for his colleague.

But Cernan was acutely aware for the rest of his life that everything he achieved on the Moon's surface was owed to the mistakes and successes of the Apollo missions before his, and of the Gemini missions before that, one of which Cernan had flown, and the Mercury missions before that. Accordingly, he recognised that, in

---

* Rovings in the Rover were inevitably limited, not just by battery power, but by oxygen supply. After all, if the thing broke down, they were going to have to walk back. Hence the carefully calculated 'Walkback Limit'. Still, Cernan and Schmitt were able to venture almost 2 miles further than anybody had ever gone before them.

order to assume its full value, his achievement and that of the Apollo programme as a whole needed to become part of a continuum of exploration. It needed to lead somewhere.

Or to quote the rocket scientist Werner von Braun: 'To make a one-night stand on the Moon and go there no more would be as senseless as building a railroad and then making only one trip from New York to Los Angeles.'

Yet Apollo 17 became a book end. Cernan's frustration was evident at a panel event he took part in for the mission's fortieth anniversary in 2012. 'We busted that door to the future open,' he told the audience, his emotion audible. 'We busted it open for those walking in our footsteps to go on through it. And here we are four decades later and we can't even get an American into space on an American piece of hardware.'

Finally, though, the story can be picked up again, the thread pulled through. At the time of writing, six Artemis flights have been proposed and the second or third of the landing missions could well contain a European astronaut, trained by the European Space Agency. Among the proposed projects is the creation of a permanent base at Shackleton Crater at the Moon's South Pole, perhaps as soon as 2030. And beyond that we can seriously begin to contemplate the launching of crewed missions to Mars, followed by the creation of safe habitats there. Mars, by the way, is around 140 million miles away. In just a handful of years we could be sending people so far that the Earth becomes no brighter than any other star in the distance.

All in all, we stand on the brink of what will almost certainly be the most adventurous and exciting decade yet in the history of spaceflight. So it seems like a good time to be familiarising ourselves with the people at the heart of that business – to be looking back and looking forward, and considering what it has taken historically to do an astronaut's work, and what it's going to take this time. In the pages that follow, I'm going to dig deep into some of the hundreds of amazing stories from spaceflight's history – stories of courage and grave risk, of the overcoming of seemingly insuperable odds, of seat-of-the-pants innovation, and just occasionally downright lunacy. I'll contemplate, for example, the consequences of sandwich smuggling on a long-haul space mission, and consider the part played by a domestic bathtub in the original emergency evacuation plans for Russian cosmonauts in the Voskhod programme. I'll recount spacewalks which went wrong and spacewalks which went spectacularly right – not least Bruce McCandless's astonishing, untethered outing from the Space Shuttle in 1984, pictured on the jacket of this book, a feat of solo daring which fills even other spacewalkers with awe. I'll explore incidents which best reveal the stresses and strains of the job, its risks and its challenges, its moments of anxiety and its passages of pure joy. And I'll use those stories to try to track the changing nature, character and motives of the astronaut, along with the changing natures, characters and motives of some of the other key figures who have made the history of spaceflight happen.

In what ways will the Artemis astronauts be the identical twins of their Apollo predecessors, and in what ways will they be different kinds of people altogether? How much has changed, over these fifty years, about the ways astronauts are trained, and about what is expected of them, both professionally and publicly? How much has altered about what are felt to be the basic qualifications for the job and the required psychological profile? Tom Wolfe's book about the Mercury astronauts, America's first official draft of space pilots, immortalised the phrase 'the right stuff' to describe the attributes needed to embrace the prospect of blasting out of the Earth's atmosphere for a while and then re-entering it at 24,000mph, with all that journey's inevitable dangers and uncertainties. That phrase has dominated the perception of the astronaut's character type ever since – perhaps misleadingly and unhelpfully from time to time. What form did that 'right stuff' actually take, back in those fabled first days of spaceflight, how did the 'right stuff' evolve over the years, and what will we consider to be the 'right stuff' this time?

Because some things, clearly, have moved on. Spaceflight in the era of Mercury, Gemini and Apollo was exclusively male and exclusively white, yet neither of those things will be true of the diverse crew that is set to fly on Artemis II – Reid Wiseman, Victor Glover, Christina Koch and Jeremy Hansen. How much credence should we still grant the stereotypical image of the US astronaut corps of the 1960s – preternaturally unflappable military pilots with buzz cuts and rangy grins, gliding around Cocoa Beach in convertible Corvettes? Equally, how much should we trust the standard image of the first draft of Russian cosmonauts, smilingly playing tennis in woollen tracksuits during their downtime? And how fully do those received images represent the subsequent Space Shuttle crews and ISS crews, and the crews who will soon be venturing to the Moon?

Within a couple of weeks of their return to Earth, Cernan and Schmitt, along with Ronald Evans, the third member of their crew, who had piloted Apollo 17's Command Module, were standing in a gale of cheers at the Super Bowl, waving from an open-top car as it processed slowly round the Los Angeles Coliseum behind a flag-draped flat-bed truck bearing the heat-seared capsule from which they had recently crawled. Like all the returning Apollo crews, they were instantly projected and celebrated as national heroes and international stars – superhumans who had faced down fear, coped with vast amounts of technical complexity under pressure, and gone further than all humankind before them.

They stood forever afterwards in the full glare of what they had achieved, and it grew hard for the rest of us to look behind that and see them as the products of their backgrounds, their working environments, of politics and day-to-day life, as people with families and flaws – as human, ultimately. So this book is partly an attempt to restore astronauts to their context, to flesh out their stories – to bring them back to Earth, in a sense, though not by any means to diminish them in the process. On the contrary, I'm convinced that, by taking this approach to the history, we arrive at an even greater and richer respect for these people and what they did.

We're now in an age of privately funded space missions, with companies such as SpaceX, Virgin Galactic, Blue Origin and Axiom Space, an age in which we more and more blithely speak of sub- and low-orbit space tourism. Down the years, among the earliest space-bound pilots – and later, as we shall see, among the Shuttle crews and ISS teams – the notion of casual participation in spaceflight with minimal training has tended to generate some dismissive (and, of course, self-protective) muttering.

Then again, when you think about it, the chief concern of most passengers on long-haul airline flights these days is probably what movie they're going to watch and when the food trolley is coming round – a prospect which might have startled and possibly even irked the pioneer aviators all those years ago who risked their lives in the air to make long-distance flight a possibility. So perhaps we are merely seeing the early stages of that story being repeated in space.

In any case, my subject in this book is not the developing business of private spaceflight with the investment of tech billionaires – as interesting as that tale is – but the astronauts whose pioneering adventures have brought space tourism into view and made it plausible in the first place.

Along the way, I'll be in the fortunate position of being able to bring some of my own experiences as a member of the global corps of flown astronauts. Even now, this far into the history of space-flight, there are not that many of us – 628 as I write. I'm also proud to be a member of the smaller group – about a third of that number – who have walked in space and who have discovered what it's like to be separated from that ultra-hostile yet strangely alluring environment by the thickness of a visor. But my main desire here is not to tell my own story, which I have done elsewhere. It's to tell the story of the astronaut species, and its evolution.

With spaceflight, the human story is where inspiration truly takes hold for the rest of us, looking on. Yes, we could choose to send an uncrewed craft to the Moon and it could get there more easily than a crewed craft and certainly more safely. It might even take better pictures when it's up there. But somehow only when a human has travelled that journey and seen it through human eyes, and come back and told us about it, human to human, do we feel a visceral connection with that journey – a connection strong enough to inspire us. It's that human connection with space that astronauts have been offering us since 1961, when Gagarin became the first human to break beyond Earth's gravity. It's that connection which the crews of the Space Shuttle and the ISS have offered in the meantime. And it's that human connection which the astronauts aboard Artemis II and beyond will once again supply for us as lunar exploration recommences, once more altering the known horizon and challenging our sense of the possible.

No disrespect to Helga, Zohar and Captain Moonikin Campos, of course, and their courageous and noble work on Artemis I, but it has always meant so much more when there were humans on board. This is the story of those humans.

CHAPTER ONE

GETTING THE JOB

'How did one get to be an astronaut? For that matter, just what the hell *was* an astronaut?'

– Eugene Cernan, *The Last Man on the Moon*

I. THE FIRST OF THEIR KIND

It's 2.00 p.m. on 9 April 1959, and seven men in a mix of grey and dark suits, two with bow-ties, are sitting in a row behind a long, narrow, cloth-covered table dotted with microphones and ashtrays, looking out calmly but slightly warily at the audience in front of them – 200 jostling news reporters and photographers packed into a building just a short walk from the White House in Washington DC. The room is overstuffed and already uncomfortably hot.

Close behind these seated men is a pleated curtain, to which has been attached the logo of America's recently formed National Aeronautics and Space Administration. Off to the side of that, hanging from a pole, is the flag of the United States, and beyond that, stuck to the wall, is a schoolroom-style poster of the Sun and Earth. A plastic model of a rocket and another, on a larger scale, of the capsule at its tip have been stood against the front of the table, where, leaning backwards slightly, they can't help but look a bit underwhelming – a decorative afterthought.

We are in Dolley Madison House, NASA's federal office. To be exact, we are in what was once this house's ballroom, where in the early nineteenth century Dolley Madison, then America's First Lady, famous for wearing turbans and pearls, threw lavish parties and dances and served her guests pink ice cream, but where now NASA's first Administrator, Dr T. Keith Glennan, stands a little stiffly at a lectern and gets this over-subscribed press conference underway.

'Ladies and gentlemen,' Glennan says, 'today we are introducing to you and the world these seven men who have been selected to begin training for orbital spaceflight.'

We are about to meet (Glennan informs the room) a group of exceptional individuals who have demonstrated 'to scientists and medics alike' their 'superb adaptability' for this unprecedented mission.

'Which of these men will be the first to orbit the Earth, I cannot tell you,' he continues, heading off the reporters' obvious first question. 'He won't know himself until the day of the flight.' Before then, two years of training lie ahead, 'during which our urgent goal is to subject these gentlemen to every stress, every unusual environment they will experience in that flight.'

And then comes the big reveal – the release to the world of information which has been guarded so closely that, in an abundance of caution, each of the successful candidates was instructed to check into their hotel the previous evening under a supplied pseudonym (the hotel manager's name), and knew nothing about the identities of the other six until this morning when they met.

'It is my pleasure to introduce to you – and I consider it a very real honour, gentlemen – from your right . . .'

As Glennan reads out their names, each man rises from his chair and stands, hands behind his back or laced self-consciously in front of him.

'Malcolm S. Carpenter . . . Leroy G. Cooper . . . John H. Glenn Jr . . . Virgil I. Grissom . . . Walter M. Schirra Jr . . . Alan B. Shepard Jr . . . Donald K. Slayton . . . the nation's Mercury astronauts.'

There is a burst of applause, and Eugene Cernan's question at the top of this chapter suddenly has an answer of sorts. What is an astronaut? *This* is an astronaut.

The questions from the press are just beginning, though, and the very first of them, called out by a reporter in the pack, seems to catch everyone at the table slightly by surprise.

'I would like to ask Lieutenant Carpenter if his wife has had anything to say about this, and/or his four children.'

Carpenter, who to the extent that he expected anything on this occasion has mostly been preparing himself for questions about the mission or perhaps about his own professional background and military record, seems a little thrown. Nevertheless he leans towards the nearest microphone and replies, 'They are all as enthusiastic about the program as I am.'

With that he sits back again.

'How about the others?' asks the reporter. 'Same question.'

'Suppose we go down the line, one, two, three, on that,' suggests Walter T. Bonney, NASA's Director of the Office of Public Information, who is coordinating the Q&A session.

'The question is, has your good lady and have your children had anything to say about this?'

Donald 'Deke' Slayton would later recall this moment in his autobiography. 'Somebody asked if our wives were behind us,' he wrote. 'Six of us said "Sure", as if that had ever been a real consideration. Glenn piped up with a damn speech about God and family and destiny. We all looked at him, then at each other.'

In fact, Lieutenant Colonel Glenn, who is one of those in a bow-tie, would get onto the subject of God and destiny – and also the place in American legend of the Wright Brothers – during some of his other answers in the conference, but not during his response to this one, which is, nonetheless, smoothly delivered.

'I don't think any of us could really go on with something like this if we didn't have pretty good backing at home, really,' Glenn says, a natural ease under public questioning immediately evident. 'My wife's attitude towards this has been the same as it has been all along through all my flying. If it is what I want to do she is behind it, and the kids are too, a hundred per cent.'

'My wife feels the same way, or of course I wouldn't be here,' says Captain Grissom. 'She is with me all the way. The boys are too little to realise what is going on yet, but I am sure they will feel the same way.'

Then it's Lieutenant Commander Schirra's turn. 'My wife has agreed that the professional opinions are mine, the career is mine, and we also have a family life that we like,' he says. 'This is part of the agreement.'

The room silently absorbs this robust declaration.

'I have no problems at home,' says Lieutenant Commander Shepard. 'My family is in complete agreement.'

Shepard's crispness generates a tension-dispersing laugh.

'I can say what the other gentlemen have said,' says Captain Slayton. 'What I do is pretty much my business, profession-wise. My wife goes along with it.'

Family support solidly confirmed, discussion moves on to the issue of cigarettes.

'I notice that the three gentlemen on our left have been smoking,' says a reporter.

Indeed, during the preamble, while Dr Glennan was hymning their 'superb adaptability', Slayton, Schirra and Shepard casually lit up. The cigarettes are still between their fingers.

'I wonder,' the reporter continues, 'what they are going to do for a cigarette when they get up there.'

Bonney, who has spent the Q&A session so far taking shallow drags on his own cigarette, turns this question over to Dr W. R. 'Randy' Lovelace, Chairman of the NASA Special Advisory Committee on Life Science and the man who has been responsible for coordinating the medical testing of the candidates.

'I think they're pretty mature men and we'll leave it up to them in large part,' Lovelace replies. 'Of course, we have a few months for an indoctrination programme . . .' he adds, to muted laughter.

In which case, what about drugs? In the course of the unprecedented flight that lies ahead of them, with its immense and even, as yet, unidentified challenges, will the gentlemen be using stimulants to combat fatigue – injections or pills?

Bonney refers this question to Brigadier General Donald D. Flickinger, Assistant Chairman of the Life Science committee.

'Specifically, no,' Flickinger replies. 'We won't resort to any pep pills . . . We will have no need to have artificial stimulation. They have their own built-in governing factor which is quite adequate.'

Blessedly for the Seven, other questions from the floor have less spin on them.

'Could we have the present home addresses of these men?'

They certainly can.

Scott Carpenter: '11911 Timmy Lane, Garden Grove, California.'

Gus Grissom: 'Presently my home is 280 Green Valley Drive, Enon, Ohio.'

Al Shepard: 'My family and I presently reside at 109 Brandon Road, Virginia Beach, Virginia . . .'

With the exception of Glenn, who has charmingly introduced himself as 'the lonesome Marine in this outfit', the group seem raw in this context, slightly uncomfortable in their civilian dress (although that could be the heat), a little exposed in the glare, definitely under-rehearsed – short of media training, we might automatically think now. When asked to make sense of space travel and his own ambitions, Slayton almost seems to shrug.

'We've gone about as far as we can on this globe,' he says, 'and we've got to start looking around a bit. We have to go somewhere and that's all that's left.'

That's all that's left: it's not the most resounding of clarion calls for a new age of exploration beyond Earth's bounds, maybe. But Slayton, the veteran of Second World War combat missions over the Balkans and Japan, had never claimed that he was born to write poetry about this mission.

Nor did he ever claim he was ready for this press conference. All that fighter jet experience behind him, and yet . . . 'I've never seen anything like it before or since,' Slayton later wrote. 'It was just a frenzy of light bulbs and questions.'

Even as that frenzy rages, the reporters are poring over the press kits they have been issued with, looking for links, connections, the common threads. All of the Mercury Seven are military test pilots. Three are from the Air Force, three from the Navy and one from the Marines. Is there anything in those proportions? It was just the way it fell out, say NASA. They are all in their thirties – ranging from Cooper, thirty-two, to Glenn, thirty-seven. They are all either eldest or only sons. Three of them are named after their fathers. They are all married with children. They are all Protestant. (The press conference will ask them all about their church-going, and all will say they regularly attend.)

And, as Brigadier General Flickinger made clear, they all have 'their own built-in governing factor' – whatever one of those is.

Also, every one of them has an IQ at least 10 per cent higher than 'any of us here' – at least, according to the Brigadier General. This detail about the super-intelligence of their husbands is, apparently, complete news to their wives, who will joke about it to each other afterwards.

Not that Rene Carpenter, Trudy Cooper, Annie Glenn, Betty Grissom, Jo Schirra, Louise Shepard and Marge Slayton are in Washington right now. They are all at home with their children, where the press will soon be streaming over their front lawns and knocking on their doors, because, whether it's their 'business' or not, and whether they like it or not, those partners are bound up in this story now and will very shortly discover how tightly.

But anyway, were there any significant patterns here? Any runes to be read? There had to be, didn't there? The reporters in that room in Washington focus in, looking for the unifying characteristics, the things that made these men what they surely must be – quintessentially American, the Wright Brothers in a new, improved, physically ideal form. They must be this, because the times demand it: America, and perhaps the entirety of the western world at this moment in history, is looking for heroes, and however they might feel about it themselves, the Mercury Seven are going to supply them.

'Could I ask for a show of hands,' asks a reporter, 'of how many are confident that they will come back from outer space?'

Each of the Seven immediately raises a hand, except for Glenn and Schirra. Those two raise both their hands.

II. THE RACE FOR THE SKIES

Carpenter, Cooper, Glenn, Grissom, Schirra, Shepard, Slayton . . . Eugene Cernan later wrote, 'Every military pilot in America wished his name was on that little list.' Was that strictly true, though? Clearly, notwithstanding the awkward questions, many military pilots will have looked on enviously at that crowded room in Washington, watching their fellow servicemen get transported from obscurity to international fame in the pop of a flashbulb.

Yet it was equally clear that there were pilots who, at this point in history, took a dimmer view of this supposedly exciting new opportunity in aviation – who remained unclear about what it entailed, unsure about the kind of flying it might involve, unsure about whether what it involved would even *qualify* as flying, let alone the kind of flying befitting the dignity of an elite military test pilot with 1,500 hours or more in his log book.

After all, hadn't American attempts to guide a rocket-launched capsule into space thus far been crewed exclusively by . . . monkeys? Was it for this that these men, the finest of their military generation, had spent years putting their lives on the line in fighter jets? To step into a place vacated by . . . monkeys?

And, even more disconcertingly, hadn't precisely none of those monkeys yet come back alive?

One pilot definitely had some reservations. Any claim implying universal awe for the Mercury Seven and their proposed mission needs to be weighed against the caustic scepticism of Chuck Yeager, military test pilot and American aviation hero – the first pilot anywhere in the world to break the sound barrier. The requirements for Mercury astronaut qualification, as set by NASA, included an engineering degree, and Yeager, who joined the Army at eighteen, wouldn't have been eligible, even supposing he had wanted to be, which it seems he did not.

At any rate, taken aside somewhere and asked – by no means for the only time – whether he was sad not to be among those entrusted to fly a Mercury capsule, Yeager shook his head.

'It doesn't really require a pilot,' he explained. 'And, besides, you'd have to sweep the monkey crap off the seat before you could sit down.'*

---

* Yeager made his record-breaking assault on the sound barrier on 14 October 1947 in a Bell X-1 experimental aircraft, at a height of 45,000 feet. Despite his

In April 1959, then, even amid the hype and hoop-la and the sense of a dawning new era for human exploration, it remained a real question: were these seven men in suits at that table in Washington swiftly bound for astonishing new heights in aviation? Or had they just signed on the line for a job that could only lead them down a short road to mockery and/or death?

Yet there was another way to frame America's first astronaut recruitment process: as a call to serve the United States in an hour of need – literally, a call-up. And that at least was something every military pilot in America, including Yeager, would have been able to recognise and respond to.

Getting a human into space had become a matter of national urgency, dramatically, on 4 October 1957, when the communist Soviet Union successfully launched the first orbital satellite, Sputnik 1. For three long and dismaying weeks – at least from an American point of view – Sputnik's silver ball and spidery antennae circled the globe, gleefully broadcasting radio signals down to Earth, until its battery power ran out and it fell back into the atmosphere and burned up.

Coolly dismissed by the White House as 'no surprise', this game-changing breakthrough nevertheless sent the United States, governors and public alike, into a froth of anxiety and self-examination. The future President, Senator Lyndon B. Johnson, wrung his hands: 'The Soviets have beaten us at our own game – daring, scientific advances in the atomic age.' The Democrat Governor of Michigan, G. Mennen Williams, was so incensed he broke out in rhyme:

---

early contempt for the 'spam in a can' Mercury guys, in 1962 he agreed to run the US Air Force's Aerospace Research Pilot School which trained astronauts for NASA.

Oh little Sputnik, flying high
With made-in-Moscow beep,
You tell the world it's a Commie sky
And Uncle Sam's asleep.

If the Russians could casually buzz America with a satellite, its radio signal hauntingly trackable on the ground from California to Long Island, with what else might they shortly be able to target American towns and cities – 'like kids dropping rocks on the cars from freeway overpasses', as Johnson put it? Had the USSR just served notice of an intention to colonise the heavens? It very much looked like it, and Johnson and Williams weren't by any means the only people in America wondering how this could have been allowed to happen.

Just a month later, the Soviet Union launched Sputnik 2, this time with a dog on board. Laika – literally 'Barker' – whom press releases romantically insisted was a stray from the streets of Moscow, became the first animal to orbit the Earth, categorically proving that a living, breathing creature could survive the trauma of a rocket launch and the effects of low gravity and increased radiation beyond the limits of the atmosphere – genuinely pioneering and useful work for which she would be garlanded with medals of honour and statues.

Not that Laika lived to tell the tale, though – and nor was there ever any expectation that she would. While the Soviet scientists had evidently leapt ahead with a method for getting heavy objects into space, they were still some way from solving the other half of the puzzle: how to bring them back. Laika had flown on a one-way ticket.

Nevertheless, the Soviet government could announce to the world another entirely successful mission, including the humane sacrifice of that heroic dog by controlled euthanasia just before her six-day supply of oxygen ran out. Not until 2002 would a contending version of events emerge in which Laika had in fact died of overheating somewhere within five and seven hours of launching. But nobody needed to be troubled with that detail at the time. The headline was: the Soviet Union was now right on the brink of getting a human into space.

They were certainly closer to it than the United States were. Thus far, American attempts to develop rockets big enough and reliable enough to transport objects beyond the reach of Earth's gravity had been stalked by failure, to the extent that the project had become a running joke – a national humiliation, even. The Atlas rocket, originally a long-range missile delivery system, on which up-scaling efforts had been concentrated, was tested three times in 1957 and only on the third of those occasions did it manage not to blow up on the launch pad. In 1958, the year after Sputnik, only eight of the Atlas rocket's fourteen test launches were successful, and some of those were deemed to be only 'partly successful', in the sense that the rocket did some of the things it was intended to do, but not all of them. Needless to say, when it came to rockets, not least ones intended eventually to bear human crews into space and bring them back again, 'partly successful about half the time' wasn't really going to cut it.

Meanwhile, the US Air Force had been scrambled to run a project called Man-In-Space-Soonest, known as MISS – surely one of the most unfortunate acronyms ever adopted by a military organisation, and certainly a barn-door target for satirists and headline writers, who could snigger about MISSed deadlines and things being a bit hit or MISS.* Nine highly experienced pilots

---

* Once MISS had successfully put a man in orbit, there was a plan for a MISS-2, using a heavier, longer-duration capsule, where MISS stood for Man-In-Space-Sophisticated. That's possibly even worse than Man-In-Space-Soonest.

at Edwards Air Force Base in California had been deployed to explore the possibility of orbital flight by all methods available. One of those pilots was a certain Neil A. Armstrong. Yet the project was cancelled after less than two months, in August 1958.

What did for MISS so quickly was not that clunky acronym, as it happened, but a decision to entirely re-strategise America's space effort. Putting a person in space was now officially a top priority, and the project was going to need more oomph than MISS alone could provide.

Even so, President Eisenhower's White House was keenly aware that, in the context of Cold War tensions and the already galloping nuclear arms race, the announcement by the United States of a newly expanded space programme, run by the Air Force, would only look, from Moscow, inflammatory. More prudent, surely, to remove this space mission from military control and entrust it instead to a separate, civilian organisation, nominally focused on science and research and thereby sending less aggressive signals.

So, the President's Science Advisory Committee was folded into the National Advisory Committee for Aeronautics, which was itself, coincidentally, originally a catch-up operation, founded during the First World War to address the fact that the US was lagging behind Europe in aircraft technology. But now, in this new blend, it became NASA, which opened for business in Washington on 1 October 1958. And naturally, NASA's immediate and primary focus was to be on getting the first humans into orbit – and returning them safely – via what now had the title 'Project Mercury'.

And, obviously, Project Mercury was going to need . . . well, what, actually, were they going to call them? NASA's Space Task Group seem to have thought quite hard about this. T. Keith Glennan considered 'astronaut' – from the Greek, meaning 'sailor of the stars' – to be the best job title to give these incoming employees.

Astronauts had been figuring in works of science fiction for at least thirty years at this point, and an English writer called Percy Greg had named a spacecraft 'Astronaut' in a novel called *Across the Zodiac* as long ago as 1880. It had a zing to it and it would be a companion for the already common pilot variant 'aeronaut'.

But Hugh Dryden, Glennan's deputy, felt that might be stretching things. Sailor of the stars? The nearest star to Earth is Proxima Centauri, which is about 4.25 light years away, or, in other words, getting on for 25.2 trillion miles. Nobody, not even in Russia, was seriously contemplating sailing a person that far any time soon.* Wouldn't 'astronaut' invite the accusation of hubris?

Dryden proposed the term 'cosmonaut' instead, which, implying 'sailor of the cosmos', seemed to him a more accurate job description (it would, of course, become the chosen term of the frequently more literal Soviet space programme). However, Glennan, and the spirit of poetic licence, carried the day. Ignoring the possible suggestion of over-reach, these fresh appointments would be astronauts, and a new category of professional explorer, and also a new category of civil servant, was born.

The next question facing NASA was: how do you recruit for a job that has never been filled or even, in the world beyond sci-fi, heard of up to now? Where do you seek your candidates for a role for which so many of the parameters have yet to be established?

Mindful that this was meant to be first and foremost a civilian project, NASA initially contemplated reaching out to the ranks of America's amateur explorers, to mountaineers and parachutists, scuba divers and potholers, hikers and wild swimmers,

---

* Logically, it would take four and a quarter years to get there, but only if you could find some way to travel at the speed of light. Otherwise you're probably looking at a journey time in the region of 40,000 years. You'd definitely want to pack snacks.

outward-bound types who were up for an outdoor challenge during their weekends and might plausibly bite NASA's hand off for a spot of Earth-orbiting if it were on offer. This was a notion with, clearly, inclusivity on its side, not to mention media-friendly romance and storytelling. 'This time last month he was an amateur kayaker – now he's going to space!'

One problem, though: with the net cast so wide, how many applications were NASA going to end up having to sift through? And what kind of elaborate system would they need to devise to check the CVs of these outdoor enthusiasts, to ensure that they had actually climbed the mountains they said they had climbed, and clambered into the potholes they said they had clambered into? How much time would this take, and how much money would it cost?

Meanwhile, there was a section of the American workforce right under NASA's nose with a promisingly appropriate skillset and attitude by training, moreover with full service and medical records to hand, including exact, independently witnessed data on their flying experience down to the last minute. For all the President's understandable Cold War sensitivities, and the appeal of diversity and romance, it was clear that it would be simpler by far to restrict the astronaut drafting process exclusively to the military, and perhaps even more narrowly (and cost-effectively) than that: to military test pilots.

On the negative side, that approach would immediately put women out of the running. Women in the US were barred at this point from military flight training, and had been so since 1944 and the end of the Women's Airforce Service Pilots programme, a wartime contingency. It would also happen to put Neil Armstrong out of the running. Armstrong could very easily have moved over to the Mercury Project from the now defunct MISS programme, but, even though he had years of active service as a naval pilot on his CV, including seventy-eight combat missions in Korea, he had left the US Navy several years previously and was now working in a civilian role.

Not just with hindsight, surely, was there room to wonder about the wisdom of an astronaut recruitment policy that excluded the entirety of the female population and Neil Armstrong. Nevertheless, convenience swung it, as convenience so often will: with the permission of the White House duly granted, Mercury astronaut recruitment would be military-only.

So NASA duly called in the service records of the 508 military test pilots then employed by the Air Force, Navy and Marine forces, and began to work through them. They had decided to seek candidates who were forty or under, who were 5ft 11in or shorter (to fit in the planned capsule), and who were 120lb or lighter. And they would offer a pretty decent salary at the level of US Civil Service Grade 12–15, which was between around $8,000 and $12,000 per year. (For comparison purposes, the average American salary in 1959 was around $5,500.)

One hundred and ten of those 508 pilots met the initial criteria on paper, and NASA began inviting that group to Washington, in batches of around thirty-five, for an informal, no-obligation, preliminary chat.

When that first set of pilots arrived, they knew merely that they were going to be briefed about a new career opportunity that might be opening up, and they had only the vaguest notion that this might have something to do with spaceflight. And, because this was definitely *not* a military project, they had been asked not to turn up in uniform, which enabled Tom Wolfe, in *The Right Stuff*, to memorably describe a Washington scene peopled by 'thirty-odd young souls wearing Robert Hall clothes that cost about a fourth as much as their watches; in the year 1959 this just had to be a bunch of military pilots trying to disguise themselves as civilians'.*

Of course, those pilots then returned to their bases and talked to their colleagues, so that the second group – which included Deke Slayton – arrived with a clearer idea about what was going on. According to Slayton, one of his colleagues put up a hand and asked whether, in this new role that NASA were proposing, they would still be flying planes.

The response from NASA's people was, more or less: 'Don't worry – we can get you excused from that.'

This was a poor reading of the room on NASA's part. These people were test pilots who by definition loved to fly, and in some cases *lived* to fly.† The idea that becoming an astronaut might actually take them out of the air or, worse, threaten their ability to log the hours they needed to keep up their flight qualifications would have been a bright red flag for many in the room. (As we will see, arguments between the early astronaut corps and their NASA bosses over access to planes and flying time would continue for some while.)

However, that lapse aside, NASA clearly felt they were getting through to these pilots. Their proposition was, essentially: would you be interested in filling a vacancy to become the first American – and, fingers crossed, the first person in history – to orbit the Earth? And would you care, along the way, to reassert

---

* Defunct since 1977, Robert Hall was a chain of American clothing warehouses that specialised in affordable outfits. Tom Wolfe, famously a wearer of bespoke suits with waistcoats and a frequenter of Manhattan's classiest shoe shops, would have been instinctively sniffy about it. Military pilots, less so. And I can confirm the military aviation profession's obsession with quality wristwatches.
† The astronaut Gordon Cooper once told a reporter, 'I get cranky if I don't fly at least three times a month.' He was seventy-one at the time.

your country's technological supremacy over the Soviet Union with all the geopolitical advantages to America that might accrue from that? And NASA sensed enthusiasm for the job.

Indeed, after the second group had been to Washington, and despite having met only sixty-nine of those on their original list of 110, NASA were satisfied that they had seen enough: they would recruit from those first sixty-nine. You have to wonder what levels of aeronautical talent lay untapped in the forty-one suitable pilots who were never called. But such is the ultimately contingent nature of recruitment, even at elite levels. NASA did some more sorting through the paperwork and then invited thirty-six of those sixty-nine pilots to come forward for further examination. Four declined, so thirty-two candidates now reported to the Lovelace Clinic in Albuquerque, New Mexico, for a week of medicals – part one of what T. Keith Glennan would refer to, at that eventual Washington press launch, as 'a long and perhaps unprecedented series of evaluations'.

He wasn't exaggerating. Here, over the course of seven and a half consecutive days and three evenings, those thirty-two candidates would undergo a physical and mental examination more thorough than any they had known before. And it's where, we can safely say, the abiding image was born of the astronaut as a person exceptionally prepared to have exploratory tubes inserted into places about their person where they possibly didn't even realise they had places.

III. TESTED TO THE LIMIT

The Lovelace Clinic was a private civilian medical centre, with generous amounts of carpeting in the public areas and comfy leatherette armchairs in the waiting room – plus, of course, as a photograph from the time clearly shows, a decent scattering of domed silver ashtrays on pedestals, which these days a healthcare setting would probably feel it could get away without, but which in those days were obviously mandatory.

The clinic was run by Dr William Randolph Lovelace II, whom we met earlier at the Mercury press conference tactfully seeing off a question about smoking, coincidentally enough, and who was a former serviceman with his own place in aviation history: in the 1930s, 'Randy' Lovelace had played a key part in developing the first high-altitude oxygen mask for pilots. He had also performed a data-gathering parachute jump from the not generally recommended height of 40,200 feet, during which he had been knocked unconscious by the shock of the parachute opening and had one of his gloves torn off, exposing his hand to frostbite. For his efforts he was awarded the Distinguished Flying Cross. NASA made Lovelace chairman of its Special Advisory Committee on Life Science and then entrusted him to run the medical testing programme at his clinic, which had already been the venue for the screening of U2 spy plane pilots so had government-approved confidentiality protocols in place.

And now through its doors came the potential future astronauts of America. For all of these pilots, medicals were a routine aspect of the working year. But not like this. 'If you didn't like doctors,' Deke Slayton wrote in his autobiography, 'it was your worst nightmare.' The candidates were examined by a general surgeon, a flight surgeon, an eye specialist, an ear, nose and throat specialist, a cardiologist and a neurologist. And that was just on day one.

They were quizzed about their medical histories and their family's medical histories. They were asked about their aviation histories: numbers of combat missions, accidents, bail-outs, uses of the ejection seat and experiences of explosive decompression – anything that could have already taken a toll on their bodies.

They had their blood screened for cholesterol, carbon dioxide, potassium, sodium, chloride, urea and protein-bound iodine, to name only those. They had their eyes tested and their extra-ocular muscle balance checked, and photographs were taken of their conjunctival and retinal vessels. They had an electrocardiogram, a ballistocardiogram (measuring the forces generated by the heart) and an electroencephalogram (a recording of their brain activity), and X-rays were made of their teeth, their sinuses, their thoraxes, their stomachs, their colons and their spines. They endured a proctosigmoidoscopy (an internal examination of the lower colon with a long instrument which the pilots immediately nicknamed 'the steel eel') and they dutifully surrendered for analysis their saliva, their throat cultures, their stools and, yes, their sperm.

What sperm motility had to do with someone's ability to pilot a spacecraft into orbit nobody really knew, but NASA were clearly determined to find out.

They had their reflexes assessed and were thrown about on a tilt-table to check the efficiency of their vasomotor control by their autonomic nervous system. They pedalled against a gradually increasing load on a bicycle ergonometer until their heart-rates reached 180bpm. They had their lung capacities and their ventilation efficiencies tested. And they had their body densities calculated by being weighed naked in water after breathing in, and again after breathing out.

When asked later which of the tests had been the worst, John Glenn replied, 'It's rather difficult to pick one, because if you figure how many openings there are on a human body and how far you can go in on any one of them . . . Now, *you* answer which one would be the toughest for *you*.'

However, all but one of the thirty-two got through this ordeal, and the one who didn't failed on what seemed to be the slightest technicality: a fractionally high bilirubin count in his blood, indicating a possible liver issue. That candidate's name was Jim Lovell, and he is possibly familiar to you because he was accepted by NASA three years later in 1962 as part of the second astronaut draft – the Next Nine, as they were labelled – and later flew on Gemini 7, Gemini 12, Apollo 8 and Apollo 13, becoming the first person to fly in space four times. So much for his bilirubin count.

The other thirty-one were handed their completed medical assessments and instructed to report with them to Wright-Patterson Air Force Base in Dayton, Ohio, for part two of this examination process: 'Stress Testing'.*

Here they were exposed to the delights of the Harvard Step Test ('subject steps up 20 inches to a platform and down, once every two seconds for five minutes') and the maximum-workload treadmill (platform elevates by one degree every minute until the subject's heart-rate reaches the 180bpm threshold). They were strapped to a steeply inclined table for twenty-five minutes while their heart was monitored to see how it would adapt to the head-down position. They tried their hands at the 'Complex Behavior Simulator' – an electric panel with twelve signals, each requiring a different response, designed to assess the subject's ability to react reliably in confusing situations. They were spun around in a human centrifuge to see exactly how much spinning and dipping

---

* It was extremely trusting of NASA to place these pilots in charge of their own medical reports – akin, as Mike Mullane puts it in his extremely colourful memoir *Riding Rockets*, to 'trusting a politician with the ballot box'. Flying down to Houston with his Air Force medical folder, ahead of the 1978 draft, Mullane took the precaution of removing a few pages which he felt didn't reflect entirely well on him, including a report on an incident of whiplash suffered during ejection from an F-111 fighter-bomber. Doctoring the doctors, I suppose you would call it. He was selected for the corps and reinserted the offending pages on the flight back.

they could take. They plunged their bare feet into ice-cold water while wearing heart and pulse monitors. They had ice-water piped into their ears to induce vertigo, and then had their recovery times measured. They were put in a dark, soundproof room for three hours to discover how they adapted to confinement and the absence of external stimuli. They spent two hours seated in a chamber heated to 130°F while medics with clipboards peered through the windows and made notes.

And then it was time for their minds to get a going over. In the psychiatric portion of the testing, the candidates were asked to respond to ink blots, tell stories suggested to them by pictures, draw a person, complete unfinished sentences, fill out a 566-item 'self-inventory' questionnaire, choose between 225 pairs of self-descriptive statements and write an essay with the treacherously open title 'Who am I?'.

Some rebelled. Given a blank sheet of paper to comment on, a candidate called Pete Conrad allegedly stared at it for a while and then pushed it back across the table to his examiner. 'It's upside down,' he said. Ironically, the wit in that response may have earned Conrad marks. What does not seem to have gained him approval was delivering a stool sample in a gift-box with a red ribbon tied around it, and leaving a full enema bag on one of the examiner's desks. His report card was marked 'not suitable for long-duration flight' and he was rejected.

Yet Conrad was persuaded by Alan Shepard to reapply for the 1962 NASA draft, was accepted, and became the third man, after Armstrong and Buzz Aldrin, to walk on the Moon. Very much suitable for long-duration flight after all, then.

In 1962, Conrad was applying alongside Michael Collins, who went on to be the Command Module Pilot on Apollo 11. Collins had seen the 'blank piece of paper' test before. He had been given it during one of his Air Force medicals. On that occasion he told the psychiatrist that he saw '19 polar bears fornicating in a snow bank'. On his report card he was graded 'hostile'.

So when that test came around again in 1962 during NASA's second draft process, Collins, who badly wanted to be an astronaut, was much more careful not to be provocative. He was quite well prepared generally, in fact. The Air Force, perhaps slightly miffed only to have earned three of the Mercury Seven slots, decided to increase their candidates' odds of selection this time by giving them all a briefing designed to help them wow the examiners.

Among the hot tips: wear knee-length socks to avoid having any bare leg showing at any time below the hems of your trousers; select a long drink, if offered, at any cocktail parties or socials, and consume it slowly; and if standing with hands on hips, ensure your thumbs are to the rear. (According to the US Air Force in 1962, only women should use the thumbs-forward posture when standing hand-on-hip.)

All of this Collins soberly absorbed and put into practice as best he could during the examinations. In addition, on the last night of testing, when some of his fellow candidates suggested celebrating their release by going out to a local strip club for 'Amateur Night', Collins sensed another test and politely declined. He went home feeling very confident that he had played a blinder.

Accordingly it was with some understandable confidence that Collins opened the envelope from NASA two weeks later – confidence which turned to dismay when he found it contained a rejection letter. He had to go through it all again, blank sheet of paper and all, in June 1963 before getting the job.*

* Michael Collins's leaving present from his Fighter Ops colleagues was a dustbin with two window-shapes cut into it, painted to resemble a Gemini capsule. Once again we witness that deep and abiding military respect for spaceflight.

Deke Slayton found it bad enough going through the medical and psychiatric testing once. It wasn't just that he found it irksome and time-consuming, it was that he found it demeaning.

'I'd flown combat missions and done operational and test flying for 17 years by that point, like just about everybody else in the process,' he wrote in his autobiography. 'The fact that I had *survived* should have told them all they needed to know about stress. What were they supposed to learn from hooking me up to an idiot machine with flashing lights? Or asking me what I saw on a blank piece of paper? Or baking me in a chamber to 130 degrees? At least by putting me in a blackout chamber they let me catch a nap.'

Still, if Slayton made his scepticism about the process clear at the time, it didn't seem to bother anyone. He was one of eighteen candidates who went forward from that 'long and perhaps unprecedented series of evaluations'. And he was still on the list when NASA eventually whittled those eighteen down to a final seven.

Seven? Well, obviously. The number seven has been felt to have uniquely satisfying and even magical properties in cultures ancient and modern. Seven days of the week, seven ages of man, seven deadly sins. Seven notes in the western musical scale, seven paths to heaven, seven halls of the underworld.

Not to mention S Club 7.

And a cat, by the way, has seven lives. Or at least it does in Iran. In the West, cats found two more from somewhere.

All in all, though, it stood to reason that NASA would want to tap into that deep well of cultural resonances by offering the world the Mercury Seven.

Except not exactly, because NASA had planned initially to end up with a team of twelve. Then they decided they would only need six. However, at the end of their whittling they found themselves with seven candidates and unable to find a good reason to cut one.

So, quite arbitrarily, the Mercury Six, who were originally going to be the Mercury Twelve, became the Mercury Seven.*

By that time, extensive background checks had been done on each of the likely candidates, each of whom were made subject to covert vetting. The testing continued, in other words, but this time without the candidates necessarily knowing. A short while before the recruitment process was completed, Betty Grissom was made aware that federal investigators had been in the neighbourhood, asking questions not just about Gus, her husband, but about her, Betty. Did she cook for her family? Proper meals? How many times a week? Was she a drinker at all? How much? Did she seem to be a good patriot? And did the marriage seem sound?†

This stage of the process represented jeopardy for Gordon Cooper in particular. Four months before his selection, Cooper's wife, Trudy, herself a military pilot, discovered his affair with another officer's wife and left him, moving out of the family home at Edwards Air Force Base in California and heading to San Diego with their two daughters. Cooper had no doubt what this development would mean for his selection if it were uncovered, as doubtless it would be. He went to see his estranged wife and explained the situation: no family home, no career as an

---

* For the avoidance of confusion, the classic John Sturges western *The Magnificent Seven* arrived on screens in 1960, a year after the Mercury Seven first appeared, although of course it was a remake of Akira Kurosawa's *Seven Samurai*, which had been around since 1954. And *Blake's 7* appeared on British television screens nineteen years after the Mercury Seven made their debut. S Club 7 had the first of their four UK number one hits a full twenty-one years after that, in 1999. The power of seven endures.

† High-level vetting continued. Ahead of the 1963 draft, Gene Cernan started getting calls from worried friends. 'Are you in some kind of trouble?' they asked. An FBI agent had been round, asking questions. 'NASA was doing a thorough background check,' Cernan wrote, 'talking to everybody from ex-girlfriends to college professors.'

astronaut. Trudy agreed to a truce and the creation of a united front. She and the children rejoined Cooper at Edwards.*

Soon after that, Carpenter, Cooper, Grissom, Glenn, Schirra, Shepard and Slayton each received the following message: 'You will soon receive orders to OP-05 in Washington in connection with a special project. Please do not discuss the matter with anyone or speculate on the purpose of the orders, as any prior identification of yourself with the project might prejudice that project.'

And shortly after that they were sitting in front of the press in Dolley Madison House and seeing their lives change irrevocably in the space of thirty minutes. By the time the next day's papers appeared, the Mercury Seven were transformed – bathed in the golden light of glory. They were 'seven superb young Americans', as the *Arizona Daily Star* breathlessly described them, and the nation agreed. Yet were they really that young? The average age of the Seven was 34.1. The Soviet draft, as we shall see, was far younger – mostly in their twenties.†

Ironically, when so much was untested and unknown, it was depth of experience that NASA had gone looking for – calm heads on slightly older shoulders. That had been the point of setting the age limit out relatively wide, at forty. Much later, Michael Collins would put it simply: 'In the early days, the environment of space was expected to produce a variety of surprises, and who

* Cooper did not overestimate the threat to his career prospects of marital discord. In 1965, Duane Graveline, an Air Force flight surgeon, was drafted into the NASA astronaut corps in an intake of six science specialists. Directly after the announcement of his selection, his wife filed for divorce. Graveline's contract was cancelled and he played no further part in the space programme.
† Given that average US male life expectancy in 1959 was seventy-four, it would have made at least mathematical sense to describe John Glenn, at thirty-seven, as middle-aged.

better to cope with them than a rough old bird who had been around the block once or twice? I think it was good reasoning.'

Indeed. But 'seven rough old birds' wasn't going to make for breathless newspaper copy like 'seven superb young Americans'. Nor would it answer to the need of the times for warriors to take the Space Race fight to the USSR. The media, and America, wanted a team of shining knights, and, whatever their own feelings about it, that's what the Mercury Seven were going to provide.

That night, after the press conference, America's new breed of explorer attended a dinner with executives from NASA, military chiefs of staff and an editor from *Life* magazine, who had bought the exclusive rights to the Mercury Seven story. They were instructed to pack up their homes and report with their wives and families to Langley, Virginia, three weeks later. And on a Monday morning, 27 April 1959, the western world's first professional astronauts went to work.

IV. A DOG'S LIFE

Moscow, late March 1961. Representatives of the international press gather at the Academy of Sciences, responding to the invitation from the Russian government to meet Zvezdochka – 'Starlet' – who will be paraded before them in triumph. For Zvezdochka has just become the first dog to orbit the Earth and return alive.

Indeed, this will be a kind of pack photo opportunity, in which Zvezdochka will graciously share the limelight with other supporting members of the USSR's burgeoning canine cosmonaut team – Strelka, Belka and Blackie.

The dogs are brought in, the shutters click, the flashbulbs explode and the reporters crane for a view.

Off to one side, and completely ignored, sits a row of anonymous men in their twenties, silently looking on at the scene. Unbeknown to all but the innermost circle of Russia's space effort, these are the USSR's trainee cosmonauts and their number includes a twenty-seven-year-old called Yuri Alekseyevich Gagarin.

At this moment, then, the world's press is in a room with the man who, in only a fortnight, will become the first person to orbit the Earth. But everybody is busy looking at dogs and nobody has the faintest idea.

The Russians had been forging ahead as fast as America feared. The USSR Council of Ministers Resolution, 'On Preparing Humans for Space Flight', had been passed on 22 May 1958, almost a year before the Mercury Seven started work. Its terms included provision for a Cosmonaut Training Centre under the auspices of the Air Force. It would be built in the outskirts of Moscow and be called Zvezdnyy Gorodok – Star City.

Meanwhile, Russia had picked a team of twenty trainee cosmonauts. Why twenty? One version of history says that Sergei Korolev, the Russian rocket designer, was asked how many cosmonauts he would need.

'How many does America have?' he asked.

'Seven,' he was told.

'Then I'll have three times as many,' he replied.

And then, after one dropped out, Korolev found himself with a corps of twenty. Given the rapaciously competitive nature of the times, it feels as plausible an explanation as any other.

The official story would be that Russia had carefully scoured the entire country in search of these future heroes, leaving no stone unturned in the quest to uncover the very finest of Russian stock for this crucial task. In fact, in common with the US, and presumably for similar reasons of efficiency, they had simply scoured the Soviet Air Force. The records of 3,000 fighter pilots below the age of thirty-five were carefully examined. There was no stipulation that the candidates should be test pilots. The key skill, in fact, at this stage of the programme was probably parachuting. In the plans for the Vostok system, the pilot would not be completing his return journey inside the capsule, he would be ejecting during the final phase of the descent and, assuming he survived that ejection, would float down to the surface separately under a canopy.

Jumping out of the spacecraft and not actually completing the return journey inside it . . . might that be thought to remove a little of the dignity from the escapade? Might it even possibly be interpreted as cheating? Certainly that aspect of the Russian method for getting a person in and out of orbit would not be confessed to until several years after Gagarin's successful flight.

No matter. Two hundred and six pilots with adequate parachuting experience moved on to the medical stage at the Central Military Scientific Aviation-Research Hospital in Moscow, where they were poked and prodded and screened and analysed, and spun and twisted and dropped and tilted, with, seemingly, a vigour and thoroughness to match that of their American rivals. Twenty-nine came through the medicals and the stress testing, and twenty of those became the corps.

Among their number was a young pilot called Valery Bykovsky who on the eve of leaving his regiment with orders to report to the Frunze Airfield north of Moscow went through the military tradition of giving away his no-longer-needed possessions to his colleagues, surrendering his radio, motorcycle and air rifle. The joke was that he certainly wouldn't need them where he was going: space. (In 1963, Bykovsky would set a new endurance record by becoming the first person to spend five days in orbit.) After a stay in quarters at Frunze, he and his colleagues were relocated to two-room apartments (for the married pilots) and one-room apartments (for the singles) near the Chkalovsky air base, also north of Moscow, and there they would stay until residential quarters were finished at Star City.

Where Russia's approach differed from America's was in creating a corps within the corps – a unit known as the Vanguard Six, who would be trained up separately for the immediate task of orbiting the Earth in the tiny, spherical Vostok capsule. So tiny was that capsule that none of the six could be taller than 5ft 7in. Alexei Leonov, whom we met earlier in this book, and who would later become the first astronaut to walk in space, missed the cut for that reason. Yuri Gagarin made it, though – he was 5ft 5in. So did Anatoly Y. Kartashov, Andriyan G. Nikolayev, Pavel R. Popovich, Gherman S. Titov and Valentin S. Varlamov.

In the character references for those pilots, sent up the command chain by the selection committee for approval, a handful of phrases appear repeatedly. 'He was not ill in 1960' is one. 'The collective respects him' is another. 'He is able to keep a military secret' is a third. Also: 'he is devoted to the cause of the Party and the Socialist Motherland'. In stark contrast to the Mercury Seven, the identities of the six were kept secret until they flew, and the names of the unflown cosmonauts only became known with the release of formerly confidential documents in the 1980s. Though there are photographs of the larger cosmonaut group, no photo of the six together is known to exist.

In March 1961, those six cosmonauts, the group-within-the-group, flew, carefully separated in three planes, from Star City to the Tyuratam test site in Kazakhstan – later known as the Baikonur Cosmodrome – to watch the launch of Zvezdochka's flight. Someone who saw them arrive would later note in their diary: 'They are all rather short, and do not look extraordinary.'

For all their supposed advances, these were still tentative times for Russian rocketry. Just five months earlier, in October 1960, an R-16 missile had blown up on the launch pad, killing seventy-four people, including the Commander of the USSR's strategic missile forces. That remains the worst rocket accident in history. But it was covered up, and would have remained so if photographs from an American spy plane hadn't revealed scorch marks on the launch pad's concrete and – still more revelatory – the shape of what was obviously a newly dug mass gravesite in the nearby village of Leninsky.

On this occasion, the cosmonauts were shown the R-7 rocket and the launch site, with its vast concrete platform and, dug out beneath it, its even more impressive flame pit, 45 metres deep, 250 metres long and 100 metres wide. As with NASA's Vehicle Assembly building, it was one of those one-off constructions which no photograph or description could quite capture – and remains so to this day. It was intended to be secret, but how do you keep an excavation that huge under wraps, even in the remote steppes of Kazakhstan? The American U2 spy planes were able to spot the Tyuratam flame pit from 13 miles up in the sky.

The six were also given their first chance to climb into the actual Vostok capsule, rather than the trainer modules they had been working in at Star City. Gagarin immediately endeared himself to the engineers by respectfully removing his shoes before entering it.

As the cosmonauts watched from a tracking station 2km away, the launch with Zvezdochka on board proceeded smoothly. Alongside the dog in the space-bound capsule was Ivan Ivanovich, the mannequin to whom I briefly alluded earlier. Ivanovich was essentially an experimental science storage facility: cavities in his suit, legs and torso were carefully packed with an array of biological samples, including mice, guinea pigs, plant seeds and vials of human blood. In other respects, though, the dummy was deemed to be eerily human, perhaps even dangerously so. The previous year, Gary Powers, the American U2 spy plane pilot, had been shot down over Russian soil and apprehended. The incident was still fresh in people's minds. Given that he was stuffed full of extremely useful samples for later analysis, nobody wanted Ivan Ivanovich getting mistaken for an enemy intruder and wrestled through the door of a police station before the rescue team could get to him, which is why the ground crew had taken the precaution of writing the word 'MAKET' ('DUMMY') on a sign and inserting it under his visor.

Both Ivanovich and Zvezdochka orbited the Earth, re-entered the atmosphere and were eventually recovered from their landing site, a snow-covered gully just outside the industrial city of Izhevsk in Udmurtia. The bad weather meant it took about three hours to reach them, in freezing temperatures. But the dog was alive, and Ivanovich had not been dragged off by the police: the mannequin and his precious biological cargo were lying in a clearing by a stand of pine trees.

By that point, the cosmonauts were well on their way back to Star City. In a little more than a fortnight, one of them would be in Ivan Ivanovich's seat.

V. PERFECT SPECIMENS

At the time of writing, in the sixty-four years since the Mercury Seven, NASA have selected twenty-one further groups of astronauts, picking 339 astronauts in total. In that time, the make-up of that corps has changed beyond recognition. Given the precedent set by those early selections, it may surprise you that only 61 per cent of those 339 astronauts came into the corps from military backgrounds. There are currently thirty-nine active US astronauts, and 41 per cent of that active corps in 2023, preparing for Artemis and the voyages ahead, are women.

In *Riding Rockets*, Mike Mullane writes about a key moment for that culture shift. Mullane, a Vietnam combat veteran, was a member of the pivotal 1978 NASA draft, a thirty-five-strong intake which included Dr Sally Ride, who would become the first American woman in space. NASA grouped that draft into three categories: Pilot Astronauts, Military Mission Specialist Astronauts and Civilian Mission Specialist Astronauts.

'Military pilots were almost always politically conservative,' Mullane writes, regarding the sea-change which swept over the NASA corps as a consequence of that draft. 'But the reign of the right ended with the large number of civilian astronauts standing on that stage . . . For the first time in history, the astronaut title was being bestowed on tree huggers, dolphin-friendly fish eaters, vegetarians and subscribers to the *New York Times*.'

As Mullane's pithy way of putting it implies, the gradual assimilation of different backgrounds and viewpoints within what had started life as substantially a military monoculture did not always happen entirely smoothly, and moments of culture clash, and the development occasionally of cliques and rifts within the workforce, are part of the evolutionary story that lies ahead of us here.

But the extent of the change itself is indisputable. In so many ways, my own application for the job of astronaut took place in another world entirely from the one occupied by the pioneers of the late fifties and sixties. Half a century on from the Mercury Seven, entirely of my own volition, I responded to an online advertisement and, in the first instance, filled in an online application form, along with almost 10,000 other people. I went through a five-round application process that took place over eight months and in two countries. I joined an organisation in which a military flying background, such as mine, was still respected but by no means essential – one in which expertise in science and engineering were equally prized. And I joined a multi-national enterprise. As a British astronaut at the European Space Agency, I trained with a German, a Dane, two Italians and a Frenchman, flew in a Russian rocket from Baikonur commanded by a veteran Russian cosmonaut, and took my place in a multi-lingual ISS crew under an American commander.

But one thing hadn't altered at all: those preliminary medicals. OK, there were no silver ashtrays in the waiting room at the clinic in Cologne where I reported for the standard week of probing during my ESA selection in 2008. But Deke Slayton and company would have recognised almost everything else about it. I ran on a treadmill while someone metred my lung capacity, and while someone else repeatedly chipped away at my earlobe for a series of blood lactate samples. I spent a lot of time walking around in one of those deeply unflattering hospital gowns that does up behind and leaves a handy aperture at the rear. And I surrendered solemnly for a proctosigmoidoscopy, before which I was issued with a preliminary cleansing enema, along with instructions to insert said enema and then, in due course, eject it. Remi Canton, a friend I made during the process, missed the bit about ejecting. The doctors found him where they had left him, lying on the bed, clinging on with all his might, and quickly pointed him in the direction of the lavatory cubicle.*

And yes, I was presented with a blank sheet of paper and asked to comment. I resisted the urge to say anything about fornicating polar bears.

Consequently I can report that, during medical week's most trying moments, I briefly shared with the earliest astronauts the

* Remi later calculated that he had held firm for twenty-three minutes, which we decided to class as a world record in that performance category. However, see Mike Mullane's book for details of some further highly competitive efforts in this area. A Canton/Mullane face-off in this event would be something to see. Well, not to see. But you know what I mean.

dark and paranoid suspicion that this whole arrangement was a mad plot to torment and humiliate me personally. I can also confess that I found myself wondering a couple of times whether the tests were themselves a test – whether what was being closely examined was not the results of the tests themselves so much as how I responded to having to do the tests in the first place.

However, from these tales of minute and painful attention to medical detail in the recruitment process arises very easily the sense that a space agency in search of astronauts is necessarily a space agency in pursuit of the physically superhuman. And much though astronauts such as myself might enjoy encouraging such a notion to be out there in the world, and to precede us into the rooms we enter – in particular our own sitting rooms and kitchens – we would have to admit, if pushed, that it's a misconception.

I looked back at the declared requirements on that ESA job advertisement. It was actually explicitly stated in the rubric that ESA weren't expecting 'top-level athletes'. Rather, they expected applicants to be 'healthy, with an age-adequate fitness level', with 20/20 vision, either naturally or using lenses, and with 'the normal range of motion and functionality in all joints'. There was no suggestion that only superheroes should bother to apply.

When the Lovelace Clinic ran every rule it could find over the Mercury candidates – and similarly when ESA's doctors ran the same rules over the candidates for my draft in 2008 – what they were looking for wasn't evidence of uncanny health. What they were seeking, on the contrary, were signs of latent abnormalities. They were guarding, in as much as they possibly could, against the emergence of any kind of medical problem that might present a complication and a hindrance to their plans for the astronaut later on – when the person in question might be on the International Space Station, for instance, and an inconveniently long way from the nearest A&E department. Yes, they needed to see all-round fitness, and a clean bill of mental health in as much as you could demonstrate one – or what Brigadier General Flickinger succinctly called, when appraising the Mercury Seven in 1959, 'a superior foundation physically and a good psychological approach to hazardous experiences'. But they weren't seeking Marvel comic-book characters, either then or now.

When the idea comes up that astronauts need to be 'perfect physical specimens', whatever that might mean, I think of Deke Slayton from the Mercury Seven who, raised on a farm, was missing the ring finger on his left hand as the consequence of an accident at the age of five with a horse-drawn hay mower. He came through his 'long and perhaps unprecedented series of evaluations' despite that, presumably because, in the end, it was about something else, something more.

But what? What, when push comes to shove, is the vital ingredient a space agency is looking for in those moments? What makes the difference when everything that can be measured is more or less equal? In the second round of the ESA selection process, I was interviewed by a four-person panel which included Jean-François Clervoy, the French astronaut. Years later I was able to ask him about what could actually shape his decision in that situation. What does a choice like that – about who goes through and who goes home – come down to?

'Quite simply,' Jean-François said, 'I would come out of that room and if I thought I wanted to fly to space with that person, then that was a plus. If I didn't want to fly to space with them, then no.'

This response pulled me up short at the time, but it contains an obvious truth. Space agencies are looking for people to ride a 300-tonne rocket and then function for some period in a ruthlessly hostile environment without a naturally occurring oxygen supply and gravity. It's understood that they'll need candidates with determination and guts, people who can not only cope under pressure but thrive under it.

But it's an unavoidable fact that those same people are also going to spend an intense time in a very small space with other people. Because that, too, is the job. For all the blood data, the bone density assessments, the lung capacity measurements and the responses to blank sheets of paper, the final decision on your suitability might come down in the end to something entirely nebulous that can't be isolated with a probe or demonstrated on a graph – something entirely mundane, although never to be under-estimated. It might come down quite simply to being someone who gives the impression of being able to rub along with people.

Or to return to that repeatedly used phrase in the official character references for the Vanguard Six, 'the collective respects him'.

Of course, selection has only ever been the beginning of the journey. Whatever they may seem to have established about themselves during the application process, astronauts have it all to prove in training. And, as we shall see in the next chapter, some aspects of astronaut training make medicals week, and every probe it can thrust at you, seem like a very short walk in a sunlit park. Sometimes even a packed Washington press conference in 1959 would be preferable.

CHAPTER TWO

GETTING READY

'Training for spaceflight is like trying to drink from a firehose.'

– Doug Wheelock, NASA astronaut

I. LANDING A BRICK

Practice, we all know, makes perfect. But what if the thing you are trying to do perfectly is not available to be practised? What makes perfect then?

It's a predicament which, repeatedly through the history of spaceflight, has beset trainee astronauts. The thing you're preparing for – flying in space – is only really available to you in full when you're actually flying in space. It puts you in the slightly frustrating position of someone who longs to get on with becoming a pianist yet lacks an essential piece of kit: the piano.

To illustrate what I mean, let's consider a moment at the beginning of the 1980s – the start of the Space Shuttle era. Here was an entirely new kind of spacecraft, a reusable vehicle intended to rocket off the launch pad, reach orbit, withstand atmospheric re-entry and then land on its wheels like a plane. But how were you meant to learn to fly one? You couldn't just slap some L-plates on the back and take it out for a spin somewhere quiet with a competent adult by your side. *Columbia*, for instance, the first fully operational Shuttle, was a 165,000lb spacecraft with wings – an unprecedented creation, propelled vertically beyond the Earth's atmosphere by solid rocket boosters and clasped to a 153.8-foot-tall expendable fuel tank. Such a vehicle does not lend itself very easily to afternoon jaunts.

Indeed, there could be no dress rehearsals, no tech run-throughs, no shakedown flights whatsoever. No work for Captain Moonikin Campos and his pals here. After a series of atmospheric test flights in the late 1970s conducted with *Enterprise* – which was constructed without engines or a heat shield, was not capable of spaceflight and was piggy-backed into the air on a Boeing 747 – the five complete Space Shuttles only flew when it was time for the Space Shuttles to fly, which is to say on 135 occasions between 1981 and 2011. And on each of those occasions, unless it was your second time in command of a Shuttle, it was your first time flying it. It was that very rare thing in the development of human spaceflight: a spacecraft which underwent its test flight with human crew on board.

By extension, then, strictly speaking the first time an astronaut found out what it was actually like to land a Space Shuttle was . . . the first time an astronaut actually landed a Space Shuttle.

A few notes on how the Space Shuttle used to land. Having performed its mission in orbit, the spacecraft would re-enter the atmosphere at five miles per second, and then plummet through the skies at a rate of descent twenty times that of a commercial airliner, dropping 28,000 feet in just over a minute, and diving at an angle between 18 and 21 degrees. Eventually, under the Commander's manual control, it would perform an unpowered airstrip landing, coming in like a glider, flaring and eventually touching down at 224mph, deploying a drogue parachute to help it decelerate to a halt.

And, as with a glider, if you got it wrong, there would be no pulling up at the last minute, no going out and round and trying again. You had no engine power for that. Once you had committed to the landing, you were landing, no second chances.

Now, obviously NASA couldn't expect someone to fly a 165,000lb brick into space and then coax it gently to the floor without *some* kind of training. So, ingeniously, in the build-up to *Columbia*'s maiden flight they came up with a practice vehicle – the Shuttle Training Aircraft, or STA.

NASA considered using a Boeing 737 for this task, but instead ending up adapting a small fleet of four Grumman Gulfstream IIs. Superficially, you would have said that little connected the Shuttle with a sleek private jet like the Gulfstream II – sleek enough for Frank Sinatra once to have owned one. However, the jet's cockpit could be reworked to resemble the Shuttle's flight deck, with the left-hand seat given over to Shuttle controls, and with the windows altered and masked to offer the same view the astronaut would see from the flight deck of the Shuttle. The STA looked like a regular Gulfstream from the outside, but under the surface its airframe had been heavily reinforced to withstand the stress of what NASA planned to put it through hour after hour in the skies above New Mexico.

Taking off from the long and broad runway of the White Sands Test Facility near El Paso, the STA would fly to 37,000 feet. Then the co-pilot, in the right-hand seat, with a traditional set of aeroplane controls in front of them, would put the engines into reverse thrust – something you might typically experience on a commercial plane just after it lands, but here done in mid-air. And then it would be over to the astronaut in the left-hand seat.

Diving at an angle seven times steeper than a commercial plane's typical descent, the STA would now take on the aerodynamic characteristics, speed and approach trajectory of a plummeting Shuttle, with the trainee astronaut able to consult the head-up display on the glass in front of them and do their best to glide the machine downwards. Just above the runway, a light would come on and the co-pilot would resume control of the aircraft and swoop it up into the air again for another run. Typically, an individual session would feature ten such rises and falls.

Only when you had done this a *thousand* times – and when you had been assigned to a mission – would you be given the chance to try doing it in the real thing. Eileen Collins, the first woman to pilot a Space Shuttle and the first woman to command a Space Shuttle mission (STS-93, on *Columbia*), described the STA as 'some of the best training we have'.

In April 1981, *Columbia* launched on its maiden flight, with two crew members on board, orbited Earth thirty-six times and flew home to make a perfect landing on a lake bed at Edwards Air Force Base in California. The Shuttle era had now officially begun – and in some style. That mission has a claim to be the boldest test flight in history and, in the circumstances, at the level of raw, gutsy flying, was arguably the greatest aviation feat of the spaceflight era since Neil Armstrong piloted the Lunar Lander to the surface of the Moon.

'Don't ask me how I knew,' the Commander of STS-1 said later. 'I just had a feeling when we started re-entering that it would go great.'

A feeling? Or training, fed by a deep well of experience? Let's just say that in command of *Columbia* when it made that first return to the runway was John Young, an astronaut who launched into space six times, the only astronaut to fly four classes of spacecraft – Gemini, both Apollo modules and ultimately the Space Shuttle – and, after STS-1, the only astronaut to have walked on the Moon *and* flown the Shuttle. (He would command the Shuttle again, too, in 1983.)

It is said that nobody had trained as long for a specific mission as Young and the pilot Bob Crippen trained for STS-1 – three years, albeit mostly because the mission's launch date kept getting put back. Still, getting ready to fly *Columbia* had involved the pair of them familiarising themselves with its twenty-two manuals, each 3 inches thick, weighing a grand total of 64lb. (Next time you're thinking of complaining about the instructions for your dishwasher . . .) And they had been charged, as pioneers must be, with bringing about the utterly unrehearsed. Still, they were astronauts, so large parts of their working lives were about finding ways to practise the unpractiseable and somehow reach an accommodation with the almost completely unknown.

Indeed, as Commander of Apollo 16 in 1972, Young had had to learn to put the Lunar Landing Vehicle down on the Moon's surface. And if he could train to land the Lunar Landing Vehicle on the Moon, he must have figured he could train to land almost anything.

II. 'COMFORTABLE FAMILIARITY' AND OTHER NICE IDEAS

The astronauts nicknamed it the Flying Bedstead. Also the Belching Spider. None of them was very fond of it. Verdicts included 'counter-intuitive', 'a hairy deal', 'a dicey training aid' and – from Chris Kraft, NASA's Operations Director – the crisp review: 'It's dangerous, damn it.' Jim Lovell said, 'It made me worry about flying it.' And Lovell, who flew on Apollo 13, doesn't seem to have worried about much.

Yet, call it what they would, and swear about it as they might, all six of the Apollo commanders who landed on the Moon had practised at some stage on a version of this hard-to-tame contraption, and Bill Anders, who flew on Apollo 8, would eventually be able to look back and declare the Lunar Landing Training Vehicle, or LLTV, 'a much under-sung hero of the Apollo programme'.

Not, though, before it had almost claimed the life of Neil Armstrong.

The incident happened right after lunch on a Monday after-noon in May 1968 – just another day of training at Ellington Air Force Base in Texas, where Armstrong, who had been an astro-naut for almost six years at this point, was preparing to fly to the Moon. Specifically, he was training for the utterly unprecedented moment when he would land on the lunar surface, a piece of flying which was, needless to say, not without its pressures and complications – and a piece of flying which was complicated to train for, too.

NASA had considered a couple of options for a system that could take an astronaut to the lunar surface and return them. There was the Direct Ascent method, in which one massive rocket would do the whole job – blasting off from Earth, flying to the Moon, lowering itself gingerly to the surface and then blasting back off again to come home. This, basically, was the way millions of us had seen it done in comic books and on the covers of sci-fi paperbacks. But it would require one hell of a rocket, a beast with 12 million pounds of thrust underneath it, and, except in sci-fi, no one had yet come close to building one of those.*

---

* The Saturn V rocket which eventually took the Apollo astronauts to the Moon generated 7.6 million pounds of thrust, some way short of the thrust envisaged for the speculative and so-called 'Nova' rocket, intended to fly all the way to the Moon and back. All these years later, the SpaceX Starship is, at the time of writing, the largest and most powerful space vehicle ever flown, generating a maximum of 16.7 million pounds of thrust. But on its maiden flight it exploded four minutes after lift-off – or rather, to use the accepted terminology, it suffered 'a rapid, unscheduled disassembly'. We've all had one of those.

Alternatively, there was Earth Orbit Rendezvous. Here, using a series of much smaller rockets, you would fire components of your Moon-bound spacecraft into orbit around the Earth, and then assemble them up there before proceeding – a kind of IKEA, flat-pack approach to lunar voyaging. The great advantage of this method was that it would need much less rocket power than the Direct Ascent method. Another bonus: you could separate the crew from the heavy cargo, sending up the humans on a smaller, safer rocket. It would also leave open the option of establishing a space station in Earth orbit whose uses might outlast an initial shot at the Moon. This was very much the most far-sighted of the options.

In the end, though, NASA chose neither of those. In July 1962, to the surprise and in some cases dismay of the US government's science advisers, they plumped instead for a third method: Lunar Orbit Rendezvous, in which a Command Module, crewed by three astronauts, would carry a detachable Lunar Landing Module into orbit above the Moon, separate from it and then continue to circle with its pilot. The LLM, containing the other two crew members, would fly on down to the lunar surface and then use its own rockets to blast off again, joining back up with the CM in Moon orbit and reuniting the crew. The LLM would then be jettisoned, its work having been done, and everyone would fly home.

The chief objection to Lunar Orbit Rendezvous was, at heart, a humane one: it was the nature of the risk inherent in attempting to rendezvous two craft while circling the Moon, as opposed to while circling the Earth. If things went wrong in Earth orbit, the craft could at least potentially fall back to Earth and its occupants be saved. If things went wrong in Moon orbit, by contrast, the crew would be left circling the Moon forever, entirely unreachable – the nightmare fate imagined for Major Tom by David Bowie in

'Space Oddity'. Understandably, that was a prospect that many in government were extremely reluctant to countenance. President Kennedy's chief science adviser, Dr Jerome Wiesner, set himself very firmly against Lunar Orbit Rendezvous.

NASA, however, argued that the degree of actual risk in Lunar Orbit Rendezvous was no greater than in either of the other methods, and possibly even smaller. True, the consequence of failure in Moon orbit was more grim, but that didn't mean the chances of failure itself were any more substantial.

But there were a couple of other things to be said for Lunar Orbit Rendezvous. Firstly, it was the only one of the three available methods that had even a glimmer of a chance of meeting the deadline President Kennedy had boldly announced to the world in 1962 – the delivery of an American astronaut to the Moon 'in this decade'. Secondly (and this may have chimed particularly loudly in Washington), Lunar Orbit Rendezvous was cheaper than anything else on the market, possibly as much as $2 billion cheaper. Those factors taken together – the speed and the cheapness – seem ultimately to have comprised an irresistible case for taking the lunar orbit route, evaporating all government objections. As so often in the realm of spaceflight – and well beyond it, of course – the simplest answer to the question 'Why?' turns out to be 'time and money'.

So the US's shot at a Moon landing was going to involve a Lunar Landing Module, an altogether new kind of craft. And this altogether new kind of craft was going to require an altogether new kind of flying, which in turn was going to require a new kind of training.

Which is where the LLTV came in.

It was pretty clear from the eventual design of the Lunar Landing Module, and the way it was intended to set down on the Moon's surface on its feet, that the closest thing to it in Earthly flying terms was a helicopter. All nine of Armstrong's 1962 astronaut draft were given helicopter training if they hadn't already had some.* Armstrong had never flown a helicopter before he joined NASA, and in November 1963, in the company of Jim Lovell, he undertook an intensive two-week qualifying course at the US Navy's helicopter training field in Pensacola, Florida.†

That meant Armstrong could now go out on his own in one of NASA's Bell H-13 light utility helicopters and practise Moon-landing-type manoeuvres. All of the Apollo commanders seem to have taken advantage of the opportunity NASA laid on for this kind of practice, and almost certainly to their benefit in some measure when it came to piloting the Lunar Landing Module. By no means coincidentally, with the upcoming Artemis Moon missions in the frame, NASA have reintroduced helicopter flying into astronaut training after a period when it was not considered essential. It's also the case, of course, that the ability to pilot a helicopter safely goes hand in hand with other key human attributes, such as reliability, generosity of spirit and high degrees of moral integrity.

Or that's what we used to say at Army helicopter school anyway.

But if you're preparing to land on the Moon, helicopter flying on its own isn't really enough. In a helicopter you can duplicate the kinds of flight path you might need to carve on an approach to the lunar surface, but ultimately the controls are not the same,

---

* The September 1962 astronaut draft is variously referred to as 'Astronaut Group 2', 'the Next Nine', 'the New Nine', and, when you're really familiar with them, 'the Nine'.

† It was while Armstrong was driving back to Houston from helicopter training – on 22 November 1963, to be precise – that he learned, over the radio, of the assassination of President Kennedy. But you probably already knew that. As the old saying goes, everybody remembers where Neil Armstrong was when Kennedy was killed.

and the vehicle will inevitably behave entirely differently from any vehicle in airless low gravity. So, back in the sixties, NASA commissioned Bell Aircraft Corporation to come up with a training vehicle that could simulate the one-sixth g environment that the astronauts would be descending into on the lunar surface. This was a tricky brief but it could at least take advantage of new research into Vertical Take-Off and Landing (VTOL) technology, which was about to result in the production of the famous Harrier Jump Jet.*

Bell's lunar training machine was a long way from the fast-jet glamour of the Harrier, however. Indeed, it looked like a couple of skips having an argument with a wheelie bin. It was essentially a pile of metal boxes 10 feet tall and sitting on four spindly aluminium legs, just as the Lunar Module eventually would. The pilot perched on top of this strange steel contraption at its leading edge, in a mocked-up version of the Lunar Module's cockpit moulded from Styrofoam.

Yet the rum appearance of this eccentric craft concealed some very smart engineering. The LLTV's vertically mounted jet engine could send the vehicle directly upwards to a maximum of 800 feet, where the pilot could then throttle back the engine to support five-sixths of the vehicle's weight, thereby approximating the conditions of lunar gravity. And then, for the ten minutes or so of flight that the craft's fuel tanks would sustain, the pilot could control the vehicle's descent and horizontal movement using a set of hydrogen peroxide thruster rockets, just as on the Lunar Landing Module.

* Some of that pioneering VTOL research was the work of a British engineer, Dr A. A. Griffith, who was the Chief Scientist at Rolls-Royce. Dr Griffith dreamed of bringing the world vertical lift-off passenger planes. But that proved impossible, which is a shame because it would have eliminated all manner of runway development squabbles around Heathrow Airport.

Those rockets, when activated, tended to envelop both the pilot and the entire vehicle in a cloud of peroxide steam – hence the Belching Spider tag. The LLTV wasn't pretty, then. But it was certainly clever: a flying vehicle that paid no attention whatsoever to the science of aerodynamics.

But, of course, it had no wings or blades, so in the event of any engine malfunction it could not be made to glide to a soft landing. It stood to reason that if the Belching Spider ever went down, it would do so quickly and most likely disastrously.

Not for nothing, then, had NASA commissioned Weber Aircraft to come up with an ejector seat for the LLTV that was capable of operating at 'zero-zero', which is to say the lowest point above the ground at which automated ejection is still survivable. On that particular Monday afternoon in 1968, Neil Armstrong would certainly have cause to be grateful to Weber for their boundary-pushing work.

The LLTV was definitely familiar to Armstrong at this point. In astronaut training, you never do anything once if it can be done multiple times, and this was the twenty-first time Armstrong had zipped up his flight suit and walked out across the tarmac runway to wrestle this extraordinary object into the air.

Yet another go on the Flying Bedstead, then.

Except not.

At first Armstrong's practice session proceeded normally. He took the craft up to around 500 feet and held it there for a while. And then he gently and steadily brought it forward and down, almost to the runway. Then he sent it back up again, vertically, and began the same approach once more – all good grounding, surely, for that moment not many months ahead when he would guide Eagle, as the Apollo 11 crew named its Lunar Lander, onto the Moon's floor.

And then things went wrong. The descending LLTV abruptly

began to lean over and veer sideways. Armstrong seemed to manage to arrest it and hold it briefly in a hover. But then it veered again, and this time tilted through at least 30 degrees onto its left side. Tugging at the controls and getting no response, Armstrong's options were already diminishing rapidly. If the craft tipped over further, he wouldn't be able to use the ejector seat because it would fire him horizontally or, worse, downwards into the ground and surely kill him. Alternatively, if he didn't eject and stayed on board, the crashing vehicle would explode and kill him.

Through his headset he heard the ground controller telling him to bail. Armstrong defiantly gave it a split-second more, then punched out.

In a sharp flash of orange flame, the rocket under his seat blasted him skywards. Below him, almost simultaneously, the stricken LLTV tipped up and dropped vertically, clattering onto the tarmac on its back and bursting into flames. Armstrong then began to follow it down, his parachute opening to slow his fall for a vital handful of seconds and causing him to drift wide of the burning wreckage in the wind before he, too, hit the ground, on the grass beside the runway.

Armstrong stood and checked himself over. He could taste blood in his mouth: at some point during this adventure, clearly, he had bitten his tongue. But, miraculously, with the exception of a few nibbles from the midges in the grass where he landed, he was otherwise intact.

'ARMSTRONG INCURRED MINOR LACERATION TO TONGUE' stated the official telegram immediately fired off to Washington from Houston to report the incident, and timing it precisely at 13.28. There was no mention of the midge-bites. There was mention, though, of the 'TOTAL LOSS' of the LLTV and some bad news for Washington about the cost: 'FIRST ESTI-MATE $1.5 MILLION'.

The telegram also reported that Armstrong ejected at an 'ESTIMATED ALTITUDE' of 200 feet. But Armstrong him-self, supported by the filmed evidence, put the actual height of the ejection at just above 50 feet – 'pretty low', as Armstrong said, with considerable understatement.

In the course of the fuller investigation which followed, it would transpire that the LLTV's thruster system had malfunc-tioned: propellant had leaked out, pressure had dissipated and the rockets had shut down, meaning that Armstrong, through no fault of his own, suddenly had no control of the machine, and no means of wresting control back. The only way was out. According to the calculations of Chris Kraft, the Operations Director, Arm-strong had avoided death by just two-fifths of a second – in other words, by the merest twitch of his finger on the eject button.

In some ways, it was what Armstrong did directly after all this that marked him out as something exceptional. Many pilots who had just survived a near-fatal ejection from a crashing test vehicle might have devoted some time, both in the immediate aftermath of the incident and beyond, to careening around the airfield and telling anyone who would listen, from immediate colleagues to the catering team down to people who just happened to be pass-ing. But that wasn't really Armstrong's style. This was not, by all accounts, a person who was easily stirred, either to emotions or to superfluous words. It was also a person who was unusually self-contained and deeply laid-back, seemingly in all areas of his life, with the possibly sole exception of when he was at the controls of an aircraft.

How laid-back was he? Well, put it this way: Armstrong tended to drive his car sitting deep and low in his seat with his left leg crossed over his right knee, for all the world as if he were in an armchair in front of the television. That's how laid-back he was.*

Returning from a late lunch and hearing news of the crash, Armstrong's astronaut colleague Al Bean went quickly to check in on him. Was he sitting with his head in his hands and breathing deeply? Was he alone in a contemplative mood and thumbing a rosary? Was he standing under a hot shower with his eyes closed? No. Bean found Armstrong in the office they shared, seated at his desk, still in his flight suit, quietly getting on with some paperwork.

Was it true what he had just heard in the corridor, Bean asked. Did he just crash?

'Yeah, I did,' Armstrong replied, evenly. 'I lost control and had to bail out of the darn thing.'

Then he returned to his papers.

'I mean, what are you going to do?' Armstrong said later, with a shrug, when he was asked about the calm with which he had reacted. 'It's one of those days when you lose a machine.'

There was another close shave involving the LLTV a few months after Armstrong's. This time it was Joseph Algranti, a test pilot, who punched out of the falling machine in the nick of time and came down thanks to Weber's miraculous intervention. That was enough for Chris Kraft, who straight away lobbied

---

* Don't try this in a non-automatic car. Actually, don't try it in an automatic one, either. Apparently, although it was relaxing for Armstrong, it was less relaxing for his passengers. James R. Hansen, in the supremely thorough *First Man: The Life of Neil A. Armstrong*, has a story about Armstrong and a group of friends and children heading out in a few cars for a day of picnicking and canoeing on Jackson Lake, and the kids clamouring to go in Armstrong's car, a convertible. However, after the drive, which involved narrow winding hill-roads and which Armstrong navigated, as ever, with his legs crossed, absolutely nobody wanted to come back with him. Armstrong drove home alone.

to have all flights in the LLTV stopped before someone got seriously hurt.*

Kraft's firmest source of opposition to this notion? Neil Armstrong.

'It's dangerous, damn it,' Kraft said, as the two of them argued.

'It's just darned good training,' insisted Armstrong.

The Flying Bedstead remained in service.

'The astronauts were adamant,' Kraft said. 'They wanted the training it offered.'

Armstrong flew the LLTV at Ellington at least another ten times after the accident. In total, he spent thirty-four hours in that clumsy machine, and landed it, by his own estimate, between fifty and sixty times during those sessions. And we can only imagine how grateful he must have been for each minute of that training when, a year and ten weeks after he had picked himself up off the grass by the runway at Ellington, he found himself coaxing Eagle down onto the Moon's surface, through clouds of dust, over large and unfamiliar rocks, with the eyes of the world on him, two varieties of alarm going off in his ears and his fuel running out. But we'll explore that seismic moment more fully later.

Armstrong had this to say about the Moon landing afterwards. 'Eagle flew very much like the LLTV . . . and the final trajectory I flew to the landing was very much like those flown in practice. That, of course, gave me a good deal of confidence – a comfortable familiarity.'

'Comfortable familiarity': Armstrong's phrase could easily stand as a definition of the ideal goal of all astronaut training – a level of tutoring so complete and so deeply absorbed that something

---

* The LLTV crashed for a third time in 1971. This time it was a pilot called Stuart Present who picked himself up off the tarmac.

utterly unexperienced until that moment nevertheless has the feel of the commonplace and the entirely expected.

But how much risk would astronauts have to be put through in pursuit of that ideal state of 'comfortable familiarity'? And how much risk would they insist on putting themselves through?

III. THE SEVEN IN THE BARREL

On 18 May 1959, three weeks after beginning work, America's first astronauts, the Mercury Seven, travel to the NASA facility at Cape Canaveral in Florida in order to witness a rocket launch. The rocket is an Atlas 7-D, which, the proposition is, will one day carry them to space, pending selection. Anticipation among the group is inevitably high. This is a glimpse of their future.

It's a night launch, so particularly spectacular for those looking on. With a thunderous roar, the rocket rises off the pad through clouds of exhaust and orange fire and, with gathering power and an awe-inspiring crackle, ascends into the blue-black sky.

A minute later the sky fills with light again as, with a far-distant crump, the rocket explodes into pieces.

Down below, Alan Shepard turns to John Glenn and says, 'Well, I'm glad they got *that* out of the way.'

Looking up to where the rocket used to be, Gus Grissom seems to speak for everyone when he says, 'Are we really going to get on top of one of those things?'

Still, at least they had their training to bury themselves in. Indeed, in those early months the Seven barely had a moment to sit still. They always seemed to be travelling – and always, incidentally, racing each other to the hotel reception desk on arrival to check in. Why the rush? Because NASA would book four rooms between the seven of them, so it stood to reason that if you could get to the desk first you could be the person who got a room to themselves and didn't have to share.

Also, you wouldn't have to share with Deke Slayton, who was apparently a terrible snorer.

At the beginning of the process, each astronaut was given an area of the Mercury project to cover as their specialist area – cockpit layout, control systems, launch missiles, communications, and so on. And then they were flown out to visit the various commercial factories where those systems were being devised or built and assembled. They all had input as test pilots, though exactly how much they were listened to at this early stage is an open question.

At the headquarters of the McDonnell Aircraft Corporation in St Louis, Missouri, where the Mercury capsules were being built, Deke Slayton pushed for the layout of the controls to be as close as possible to an aircraft's dashboard, and to feature pedals as well as a stick. He didn't get his way on the latter: it was stick only. But he also pushed successfully for the capsule to be referred to officially as a 'spacecraft' – a small victory for astronaut dignity and self-esteem at a sensitive moment.

Meanwhile, on arrival at the Convair aviation factory in San Diego, where the Atlas rocket was being worked on, Gus Grissom was asked if he would mind getting up and saying a few words to the gathered staff – perhaps something motivational.

Grissom was never comfortable with public speaking, and certainly had nothing prepared. But he could hardly refuse. So when the moment arrived, and the Convair executive had finished introducing him, he stood up, cleared his throat, and did as requested.

'Do good work,' Grissom said.

And that was the end of his speech.

Well, it was heartfelt, and as good an instruction as any other. The staff certainly seemed to think so. They made signs with those three words on and put them up around the factory.

And then there was the training, which also involved a lot of travelling and which was transformative for all of them. They had formerly been people who flew aircraft; they would now become people who flew nose cones off the top of missiles. And just to complicate matters, nobody had ever done that. The change of gear was dramatic, both physically and mentally.

They attended lectures on astronomy, on spacecraft engineering and on the Mercury system. They were given their first brief tastes of weightlessness on parabolic flights in the back of C-131 Samaritan aircraft, knowingly referred to as the 'Vomit Comet'.* They travelled to Johnsville, Pennsylvania, to endure sessions of high-speed spinning in the centrifuge at the Naval Air Development Center, and to learn breathing techniques that would help them withstand acceleration forces above 6 g.

And they flew to Cleveland, to what was then called the Lewis Research Center, to be strapped into the Multi-Axis Space Test Inertia Facility (MASTIF), also known as the 'Gimbal Rig'. Here, as if on a particularly evil fairground ride, each of them was spun on three axes at a rate of anything from two to fifty rotations per minute, while they did their best to stabilise the rig by triggering small nitrogen thrusters. Each of the Seven was expected to log a minimum of five hours attempting to tame this contraption. And thus it was hoped they would learn to bring a spinning spacecraft under control, should they ever need to.†

---

* The scenes of weightlessness in the movie *Apollo 13* were obtained by this method, in the simulated zero-g atmosphere created by an aeroplane repeatedly climbing and plunging. The director Ron Howard allegedly leased an aircraft for six months to get the shots he wanted.
† As we will see later, Neil Armstrong in particular would have cause to be grateful for this aspect of his training, during the Gemini phase of his career. Incidentally, as John Glenn from the Mercury Seven sat in the Gimbal Rig, did it even remotely occur to him that one day (in 1999, to be exact) the Lewis

They also went to Little Creek, Virginia, to the Navy Amphibious Base to learn underwater egress – techniques for exiting a sunken vehicle. They would need to have this skill in their lockers because the Mercury capsule would, if things went to plan, be splashing down in water, and there was always a chance it might not float.*

This was the moment when Deke Slayton had to confess that he couldn't swim. Somehow, during the elaborate examinations prior to his appointment, this hadn't come up. Or if it had, nobody had considered it important. Slayton now quickly learned.

They trained for the possibility of an uncomfortable landing on dry land too, doing a session of survival training. This involved getting deposited by helicopter in the desert in Nevada and being left there for four days with little but their wits and a mock-up of a crashed Mercury capsule with its parachute still attached (very useful for shelter building). No doubt this was arduous, though at least some aspects of the experience would have been familiar to the Seven from their military test pilot training.

On less familiar terrain, they attended a fitting session for their silver pressure suits, an occasion which NASA converted – like so much of the Seven's training, in fact – into a photo opportunity.

Research Center would be renamed the John H. Glenn Research Center at Lewis Field? Actually, from what we can intimate about Glenn and his self-confidence, it possibly did.

* The American space programme in the sixties favoured splashdowns because it couldn't fit a big enough parachute to the nose cone to make landing on the ground reliably comfortable. The Soviets, as we saw, initially got round that problem by getting the astronaut to bail out, but they eventually worked out how to parachute the capsule back to the ground – which is why, on my return from the ISS, I came to understand the full meaning of the expression 'back to Earth with a bump'.

The famous pictures from this fitting session document the only occasion when all of the seven were suited up together.*

And during all of this, they were repeatedly peeled off individually, on a rota basis, to do week-long sessions of media interviews, appearances and talks around the country at the behest of NASA's Public Affairs Office. Now, that really *was* a trial. Amid all the other dangers, discomforts and downright indignities of training, this was the thing the Seven seem to have resisted the most, referring to it darkly between themselves as the 'week in the barrel'.

By all accounts, the Seven experienced only one major group-wide slump in morale, and that was about three months in, when it became apparent to them that, what with the centrifuge sessions and the hours in the Gimbal Rig and the survival training and the talking to the people at McDonnell and all the rest of it, they weren't getting any flying done. NASA eventually heard their pleas and got them some T-33 jets to use, then, a little later, some F-102s. Morale promptly rose again.

Gordon 'Gordo' Cooper once used a NASA jet to fly himself to a meeting. Unfortunately the nearest military airport to his destination had a runway which was technically too small to permit the plane he was flying to land. That didn't bother Cooper, though. He landed the jet anyway and then asked to have it refuelled while he attended the meeting.

When he returned to the plane, he found a Lieutenant standing by it with a dark expression on his face. Furious with Cooper

---

* The equivalent moment for the Soviet Vanguard Six ended up having an element of embarrassment about it. On the day of the fitting, only three finished suits seemed to have come back from the makers. While Gagarin, Titov and replacement (for Anatoly Kartashov) Grigory Nelyubov squeezed themselves into their new outfits, the other three stood to one side, looking sheepish.

for breaking the landing rules, the Lieutenant informed Cooper his request for a refuel had been denied.

Cooper considered the problem. And then he climbed back in the jet, took off again, and used the nine minutes of fuel that were remaining to him to fly 200 miles to another base where he could get fuel without anybody standing in his way.

Heading back home to Houston from flights in the early evening, Slayton would occasionally do a low pass over his house for the benefit of his wife, Marge, and their son, Kent, who would be out in the street playing with the other kids after school. One day, just to give the show an extra kick, Slayton lit up the jet's afterburner as he passed. The roar and the rush of air shattered two windows and sent a pack of small children rushing screaming for their homes.

Life for the Seven was hectic, clearly, and highly pressured, but also comfortable. Six of them moved into the newly built Houston suburb of Timber Cove, taking spacious single-storey houses with front lawns and back yards. (Alan Shepard chose to live in an apartment in Hermann Park, in a building which has since become a Marriott hotel.) Their splits of the exclusive deal with *Life* magazine were paying them half as much again as the salaries they had been earning in the military. They could also, if they wished, take advantage of the American General Mortgage company's new special cut-price mortgage rate for astronauts – 4 per cent.

Then there were the free cars. Well, not quite free. General Motors handed the astronauts new Chevrolets, allowed them to put 3,000 miles on their clocks and then offered them the choice between handing the car back or buying it for a knock-down price. So, a sweet deal, if not the gift it was often portrayed as.

Six of the Seven took GM up on their offer and began blatting about the place in sporty Chevrolet Corvettes. The one who didn't, perhaps predictably, was John Glenn, who continued to drive his NSU Prinz, a little rear-engined German model that – to put this politely – emphasised practicality over style. When Tom Wolfe mistakenly wrote in *The Right Stuff* that Glenn had driven a Peugeot, Glenn wrote him a letter to put him right. This was in 1979. 'Dear Tom,' it began. 'Just for your future records, and reprint corrections later on, the car was a PRINZ not a Peugeot. The Prinz was a little 2 cylinder car by BMW that got about 35–40 mpg . . . With my kids being a little older than some of the others at that time, I was already concerned about their college $ [*sic*], and the Prinz seemed like a good idea.'

Wolfe's interest in hearing about the economy figures on one of John Glenn's old cars is something about which we can only speculate. Not that Glenn sounded all that fond of the car himself. 'Don't believe the Prinz is even in production now,' he wrapped up. 'Haven't seen one for years – and good riddance.'*

Anyway, for an astronaut in training in the early sixties, with a nice house to live in and a nice car to drive and a jet to take up every now and again, things must have seemed pretty sweet, all in all.

Now, if only the rockets could be made to work.

In November 1960, with the Seven just over a year and a half into their training, NASA staged another rocket test at Cape Canaveral. This time it was the first try-out for the Mercury-Redstone pairing: a Redstone rocket under a Mercury capsule, on this occasion uncrewed.

What could possibly go wrong?

* The letter is in the digital collection of the New York Public Library: https:// digitalcollections.nypl.org/items/bc3ed860-6391-0132-efce-58d385a7b928. Wolfe corrected the error and the Peugeot is a Prinz in all later editions of *The Right Stuff*. For the record, at no point during my training as an ESA astronaut was I offered a free sports car or a favourable mortgage. Different times, alas.

At lift-off, amid the usual rumble, the rocket rose a few inches off the pad . . . then seemed to think better of it and lowered itself back down again. As it resettled, there was an almighty bang and a fizz, and the escape tower fired and blasted itself into the sky. Then came another loud crack, and up at the tip of the rocket the nose cone on the Mercury flew off, producing a puff of radar chaff like some kind of party popper, and also expelling the parachute, which flapped and billowed forlornly down the side of the rocket.

And now everyone was in a proper panic because if the wind got inside that parachute and tugged the fuel-filled rocket over, then who knew what kind of conflagration might occur?

Meanwhile, several hundred yards away, above a thankfully deserted strip of Florida beach, the escape tower completed its pioneering flight, which had taken it up to the definitely impressive height of around 1,300 feet, and arrowed back to the ground, sticking itself upright in the sand.

The rocket remained standing on the pad. Mindful of the time-honoured advice about not returning to a lit firework, NASA left the booster overnight to cool and then tentatively returned to it in the morning to drain it of its ethanol, make the pad safe and assess the damage.

Still a bit of work to do, clearly. The launch of MR-1 would go down witheringly in spaceflight history as 'The Four-Inch Flight'.

Training rockets was proving to be harder than training astronauts.

IV. KNOWN UNKNOWNS AND UNKNOWN UNKNOWNS

We do well to remember how little was known, in those early days of spaceflight, about what was out there – the extent to which science was in the dark. Our modern knowledge and assumptions have been bequeathed to us by more than fifty years of space-flight, during which time people have gone into space and travelled to the Moon, and come back and told us about it. Yet of course, time and again the nascent space programme found itself up against questions to which, at that moment in history, the only reasonable answer was, 'You know what? We just don't know.'

For example, what was going to happen to humans inside a launched spacecraft? The effects on those humans of the launch itself – well, even though nobody had sat on top of a rocket yet, those impacts could more or less be calculated and assessed. But once those humans got out there, what then? What disruption would the zero gravity atmosphere of space wreak on the intricate components of their bodies?

For instance, how were your heartbeat and your circulation going to cope? What if the valves which stop your blood from simply following gravity and pooling at your lowest point no longer functioned under the conditions of weightlessness? These were very real considerations.

Or what if the first human being got into space and then discovered that, up there, human beings can't swallow or digest anything? What would happen then? Would all of the delicately organised processes that keep us alive here on Earth still work as they needed to beyond gravity?

And then what about the psychological impact? How could you begin to predict that? What about those worrying studies which suggested that the sensory deprivation that would be experienced in space would play utter havoc with the mind, inducing hallucinations, confusions, wild mood swings . . . Frankly, would rational thought even be possible in a spacecraft? Wouldn't astronauts all go stark raving bonkers within a few hours of launch?

Then, in the late sixties, when the mission's reach expanded, there was the extent to which science was in the dark about the

Moon. What did anyone really know about this place where those astronauts were headed? The lunar surface would be dusty, that much was understood. But how dusty? After all, the dust had had a long time to gather, hadn't it? Some 4.5 billion years had gone by without anybody so much as flapping a duster around the place, let alone Hoovering.

So what if the Moon's dust layer wasn't just a light topping where the first Lunar Landing Module ended up setting down, but actually a thick drift of dust several feet deep – a dust dune? What if the craft which Armstrong had been so assiduously training to operate ultimately descended, in full view of the anxiously watching world, set its feet down on the Moon's eerily glowing surface, and then simply sank and disappeared below the dust, taking Armstrong with it? Crazy as it may sound, this slapstick, Laurel and Hardy-style conclusion to America's decade-long, billion-dollar space effort could not be entirely and definitively ruled out until someone had been there and found out.*

Or what if it was actually raining meteorites up there, such that any craft approaching the Moon's surface would find itself subject to a constant and possibly ruinous pebble-dashing?

Or what if static electricity generated by the Lunar Landing Module attracted metallic particles in the Moon dust which then stuck to the windows like bugs, or, worse, clogged the thrusters and killed the power? Or what if the pure metallic elements firmly believed to be in that lunar soil got picked up by the astronauts'

---

* In a famous misunderstanding, Buzz Aldrin recalled that, after he had descended from the LLM and joined Armstrong on the Moon, Armstrong had greeted him with 'Isn't it fun?' It didn't sound like a particularly Armstrong-like reaction. And indeed, Armstrong later maintained that what he had actually said to Aldrin at that point was 'Isn't it fine?' – or possibly even just 'Fine' – referring to the condition of the dust under their feet, and possibly expressing some relief about its depth.

boots and then tramped back into the Lunar Module; and what if those elements then reacted with the module's pure oxygen environment, leading to an instant and terminal conflagration?

And what were they meant to do about that? Pack a doormat?

And then what if the Moon turned out to be impregnated with destructive alien organisms, but we didn't know until the astronauts came back, dragging them home on their clothes and the soles of their shoes and in their rock samples, and thereby unleashing a catastrophic pandemic upon the Earth – a kind of Moon Covid?

A special NASA panel was charged with considering such things – the Interagency Committee on Back Contamination. Indeed, a genuine fear of the existence of human-hostile lunar contaminants would lead to the National Academy of Sciences recommending quarantine periods for the first Moon crews and everything they brought back with them – for three weeks in the case of the Apollo 11 astronauts. Armstrong's, Collins's and Aldrin's reward for enduring just over four days beyond the bounds of Earth was a further twenty-one days of solitary confinement, one of which was Armstrong's thirty-ninth birthday. The three of them celebrated slightly mournfully with a cake behind glass in their specially adapted Airstream trailer. Post-mission quarantine periods were only cancelled after Apollo 14, the third of the six flights to reach the Moon.

There were a million 'What ifs?', clearly. And educated surmises could take you a very long way. But the inescapable fact was, none of these questions would have a definitive answer until somebody went up there and came back.

Or, as it may be, didn't come back.

In this debate, the scientists in the organisation were generally on the side of safety and proceeding carefully, by tiny increments. And that was fairly likely to set them in opposition to the trainee astronauts, who, it's safe to say, were more broadly on the side of getting out there and seeing what happened. Consequently, the rift that was first detectable during the selection process – fussily exact scientist v. self-assured and somewhat impatient pilot – solidified during training into a feature of astronaut life.

All those blood tests and capacity measurements during the application process turned out to prefigure their lives in the job. The Seven, and all who followed them into the astronaut business, were now subject to constant physical monitoring and data harvesting – before they flew, while they flew, after they flew. This has remained true for everyone who has taken up the role since, but it must have very quickly dawned on those first astronauts that they were now, like it or not, a living experiment, the property of science.

Possibly even harder to reckon with: they were now actually *scientists*. During any space mission, the astronauts would automatically become the conduits for the information the science community back on Earth was desperate to glean. They needed to be in a position to understand what they were looking for and what they were looking at. And that meant they would need to know their way around not just engineering but, at the bare minimum, astronomy, meteorology, geography and physics. When the Moon came into reach, they would need geology too. The Moonwalkers were going to have to identify, analyse and collect rock samples, and no one was suggesting they should just get up there with a trowel and improvise. The NASA training programme in the 1960s put all of the Apollo astronauts through a geology course which involved a minimum of 300 hours of classwork and sixteen field trips.

And incidentally, just as NASA has reintroduced helicopter flying, so ESA have recently added a geology module to their astronaut training programme, the return to the Moon with Artemis driving a return to the ways of the sixties.

But as the reach of the mission expanded, so, inevitably, did the nature and demands of the job. Little wonder that NASA broadened the selection criteria for the second astronaut draft – Armstrong's draft, the team initially intended to fly the Gemini programme and then crew the Apollo missions – to make a biological sciences degree count as much as an aeronautical or engineering degree. And little wonder that in 1965 they organised a specifically science-oriented draft, inviting applications from people with doctorates in the natural sciences and medicine, as well as engineering, and bringing in six research scientists, including the geology Ph.D. Harrison Schmitt.

If the prospect of working alongside these scientists, and even ultimately flying with them as equals, triggered some instinctive resistance among the hardcore military test pilots in the corps, it was possibly even harder for those astronauts to consider that they had themselves crossed over to the scientists' side. As if the comparisons with monkeys weren't enough for a proud pilot to endure.

Of his training for the Gemini programme, Gene Cernan wrote: 'We all had highly technical backgrounds, but damn it, we were pilots! It was an image thing, and we didn't want to be known as scientists.' The endless geological field trips seemed puzzling to him. Dumb rocks, for heaven's sake! 'I crunched through the lava beds,' he wrote, 'and wondered what all this could possibly have to do with flying in space.'

Strains inevitably emerged. In the absence of irrefutable evidence, the scientists were initially extremely reluctant to push the limits on how much weightlessness they believed the human body could take. When Alan Shepard eventually became the first American to fly in space, and came back alive and undamaged after fifteen minutes, the restrictions could lift – but only a bit. The way the scientists saw it, a quarter of an hour in space was now demonstrably achievable without dire consequences for the supply of blood to an astronaut's brain, and seemingly without destructive consequences for his sanity. But did it necessarily follow logically that *whole days* in space weren't going to produce problems? Back to the drawing-board – or rather, the lab – on that one.

For some of the astronauts, this extreme caution and reluctance to let rip was only irritating. Michael Collins wrote: 'Living and working with these people was like having . . . a close friend who sincerely believes in astrology and can't stop talking about it, especially delighting in reading you your horoscope on bad days. None of it is to be believed, but it's pretty difficult to ignore.' Collins was ungrateful altogether for the medics: 'The truth of the matter is that the space program would be precisely where it is today [he wrote this in 1974] had medical participation in it been zero, or perhaps it would be even a little bit ahead, because we could have done without all the impedimenta and medical clap-trap, such as blood-pressure cuffs, exercise ergonometers, and urine-output measuring devices. All they did was add weight, and complexity, and rob time and energy from tasks of greater value.'

One comes back to Cernan, who 'wondered what all this could possibly have to do with flying in space'. That thought seems to have been frequently in the minds of the first astronauts as they trained – and has probably been in the minds, at some point, of every generation of trainee astronauts since. How much of this is relevant? How much of this is strictly necessary?

And what better way to bring those questions into focus than with a bout of extreme-weather survival training? Just like the Mercury Seven before them, Cernan and Al Bean were sent off to the desert of northern Nevada, dropped off by helicopter in the screaming heat with limited rations and given five days to make their way to a given meeting place. Did this acquaint them with any skills they eventually used on their various distinguished missions? Difficult to conclude, but it certainly acquainted them with a rattlesnake, which wriggled into their camp one evening. They were so hungry at that point that they went after it with a machete. After an energetic chase, the rattlesnake escaped, leaving them even more hungry.

Michael Collins, after being airlifted to the desert near Reno and also to the jungle in Panama, concluded flatly that nothing he had done out there in any way helped his performance in space. But of course, Apollo 11 splashed down according to plan, in the Pacific Ocean, southwest of Honolulu and just 13 miles from its recovery ship. Collins might have felt differently if things had gone awry and he'd landed somewhere remote – in the desert near Reno, say, or the jungles of Panama.

Contemporary Russian cosmonauts do a training exercise in which they are asked to solve logic tasks while freefall parachuting. Now, you could argue that there is something wildly contrived about that, just as wildly contrived as being sent into the desert to pit yourself against rattlesnakes. Parachuting and logic tasks? It's almost as if someone had been set to devise a training programme and had started shouting the first things that came into their head. 'I know: skydiving and Sudoku!'

Yet you can't deny that it would be a challenge. And that's clearly the point: the promotion of a mindset which responds well to challenges wherever they occur and whatever form they take.

Of course, the open-endedness of astronaut training could be regarded as an opportunity in itself. I certainly found myself thinking of it that way. Indeed, it was a consoling position to adopt during periods when my chances of ever getting a mission and ultimately flying in space looked remote. There are courses that will teach you Russian; there are courses that will teach you how to perform a dental extraction; there may even be courses that will teach you how to catch, kill, pluck, cook and eat a chicken while stranded on a mountain in Sardinia. But there can't be many courses that will make the effort to teach you all of those things, as my ESA training did. I convinced myself that the training itself was an experience to take away, whatever use I eventually ended up putting it to.

However, there was one aspect of training whose absolute relevance was not up for debate, and that was the time spent in the simulators. The remote space analogue environments I was fortunate enough to participate in – the Sardinian underground cave network in which I acted as consultant for ESA's CAVES programme, the undersea Aquarius habitat off Key Largo where I spent twelve days 25 metres down on the ocean floor as the first European astronaut to be selected for one of NASA's extreme environment NEEMO missions – most certainly had their excitements.* But they could offer nothing like the surge of anticipation I felt when I finally reported to the Johnson Space Center for mission-specific training on the full-size practice modules – in the Space Vehicle Mock-Up Facility, where full-size models of the ISS's various compartments were laid out on the floor; and in the Neutral Buoyancy Facility, where further full-size modules are sunk to the bottom of a 6.2 million-gallon pool, enabling you to experience working on them while afloat, and to train for space-walking in conditions akin to weightlessness. Those were the days when your mission felt close enough to touch.

* They also had their dangers. On my NEEMO 16 mission, during an exercise in the water outside the habitat, my Japanese astronaut colleague Kimiya Yui got snagged on the umbilical of a support diver who, returning to the surface, unwittingly began to drag Kimiya with him, which could have been fatal. Fortunately the support diver realised in time. And in Sardinia, soon after astronauts had completed their ESA caving course, Luigi Mereu, a local thirty-two-year-old speleologist, fell to his death while transferring equipment out of the caves. The risks were real.

No doubt the Shuttle astronauts felt the same about the two Shuttle Mission Simulators which had occupied the same building – the fixed-base SMS, which was used for orbit simulations, and the motion-base SMS, which had six hydraulic legs and could tip and shake the cockpit around to simulate launch and re-entry.

And no doubt the Apollo astronauts felt similarly about their various Lunar Module simulators, and Command Module simulators, which John Young christened the 'Great Train Wreck' because of the mash-up of boxy compartments that were stacked up around its exact replication of the Command Module's interior. These sixties and early seventies devices may not have looked sophisticated, but in fact they were ahead of their time. They were, according to Armstrong, 'the best, most advanced simulators ever constructed anywhere'. They were dynamic and interactive, and what the trainee saw from the windows during the flights was a rough approximation of the sky as it would appear – very primitive and sketchy, no doubt, by modern standards, but accurate enough to navigate by. And they brought the mission close.

Armstrong seems to have taken his own approach to handling the various emergencies that would be thrown at him during flight simulations. Even when the order to abort came in from the simulated Mission Control, he evidently liked to hang on in there, beyond the point of no return, just to see what might happen – and just to lob the management team outside the simulator a few surprises as well, thereby keeping them on their toes. The way Armstrong saw it, if you pressed 'abort', that was playing it safe – exercise over. Whereas if you carried on and eventually crashed, no one actually got hurt, and you might learn something useful for later.

This attitude does not always seem to have endeared him to his colleague alongside him in the LMS, Buzz Aldrin, who adhered to the more conventional notion that the whole point of simulator exercises was *not* to crash. According to some accounts, Aldrin and Armstrong had their only personal bust-up during a late-night discussion of the wisdom or otherwise of Armstrong's deliberate abort-avoidance during that day's simulator session.

Still, Armstrong spent 383 hours training in the Lunar Landing Module simulator and 164 hours in the Command and Service Module simulator. When you add the thirty-four hours he spent in the devilish LLTV, it means he went into the Apollo 11 launch with a total of 581 hours of simulated mission time under his belt – which, as James R. Hansen points out in *First Man*, is the equivalent of more than ten whole weeks of eight-hour days.

Just for the record, Buzz Aldrin did even more than that – forty-six hours more.

Similarly, before he became the first American to orbit the Earth, in February 1962, John Glenn had spent nearly sixty hours in the simulator, had flown a simulated version of his mission seventy times and had responded to 189 simulated system failures.

And of course, before John Young landed Space Shuttle *Columbia* for the first time he had made those 1,000 mock landings in the STA and logged at least 300 further hours of simulated trouble-shooting.

Clearly the idea of astronaut training was to try and cover every angle, account for every possibility, prepare for everything.

The problem was that you never could.

V. THE CHAMBER OF SILENCE

It may surprise you to know that there was no formal fitness programme for American astronauts during the Mercury, Gemini and

Apollo years, no dedicated slot on the training schedule for running or gym work. For all that NASA placed exact figures on the hours they were expecting their trainees to spend in the MASTIF, the centrifuge, the classroom and all the other places, when it came to physical exercise the astronauts did as much or as little as they chose.

Which, in Neil Armstrong's case, was exactly none. Armstrong once discussed with a friend an idea he was partial to, that humans had a finite number of heartbeats in a lifetime. 'I don't intend to waste any of mine running around doing exercises,' he said.

Well, it's a theory.* John Glenn seems to have taken himself out for runs on the beach quite regularly. Al Shepard took up a morning jog after he quit smoking. But they seem to have been unusual.

Contrast the Soviet cosmonauts, who were photographed running shirtless through icy pine forests and flashing hearty smiles while knocking tennis balls back and forth. Like more recent photographs of Vladimir Putin topless on horseback, these images possibly served a function as propaganda, of course. Yet it's surely a significant fact that the standard training-wear of the Russian cosmonaut was not a flight suit, but a woollen tracksuit. While the Americans were driving their Corvettes with the hoods down and swanning in and out of the motel bars on Cocoa Beach, their rivals in Russia were seemingly behaving like a sports team on a particularly intense pre-season retreat.

The cosmonauts played football, tennis and badminton. Neil Armstrong would presumably have been horrified. They also

---

* Someone else who evidently holds this view: Donald J. Trump. The forty-fifth President of the United States once expressed his belief that the human body ran on a finite battery, which exercise only depleted. And that's why you never see Trump out for a run in your local park. Among other reasons.

played chess (though not Yuri Gagarin, apparently, who didn't like the game), and cards. And they watched movies.

But of course (and here maybe Armstrong had a point), all this healthy exercise was also risky. On a day off, the cosmonauts Valentin Varlamov, who was a member of the prioritised group-within-a-group the Vanguard Six, and Valery Bykovsky, who wasn't, went for a swim in a nearby lake. Bykovsky was first to dive in and lightly grazed his head on the lake bed. He surfaced and warned Varlamov that the water wasn't as deep as it looked. Varlamov ignored him, plunged in head first and hit the bottom, injuring his neck. He had to leave the Six. He was replaced, ironically, by Bykovsky, who could at least say he had warned him.

As if training itself wasn't dangerous enough. Another member of the Six, Anatoly Kartashov, was injured while being put through his paces in the centrifuge. He was spun so hard that blood vessels ruptured in his back. Despite pleading to continue, he was stood down and replaced in the Six by Grigory Nelyubov.

And then there was the fate of poor Valentin Bondarenko.

At twenty-three, Bondarenko was the youngest of the first draft of cosmonauts, yet he was already married, with a five-year-old son, Alexandre. But his energies were boyish. He played tennis, and sang. There are tales of him running around the Star City compound, knocking up colleagues to see if they would come out and play football with him. The other cosmonauts affectionately called him Tinkerbell.

In its necessarily speculative attempts to prepare for the unknown demands of space, the Russian programme had devised the 'Chamber of Silence'. This was a small, lead-lined, sound-proof room with the barest of fittings: a metal bed, a chair, a table, a toilet, a waste bin. It also contained a single hotplate and a solitary saucepan. A prison cell might be an apt term for it, apart from the fact that this particular cell was hermetically sealed and the air in it was artificially mixed with a high oxygen content to duplicate conditions inside a spacecraft.

Each cosmonaut in turn would enter the Chamber of Silence for a period of solitary confinement; they were expected to last fifteen days behind its 16-inch-thick door. Nothing much happened in those fifteen days. At certain moments a light would come on, and scientists would instruct the occupant, over a Tannoy, to attach a set of bio-sensors to his body for monitoring. Four hours or so later, he would receive the instruction to remove them. At other moments, loud classical music would abruptly be piped into the room. The cosmonaut might have access to a pen and some paper, possibly some wood and a knife for whittling, perhaps some crosswords or number puzzles. Otherwise time was formless, and that was the test: how would each of these people cope with their own company, minimal control over their immediate environment and a profound degree of sensory deprivation over a period of slightly more than a fortnight?

Wouldn't anybody who could come through an experience like that automatically be a safe bet for space?

In March 1961, Bondarenko became the seventeenth of the twenty cosmonauts to report for this endurance test. So far as we know, all had thus far emerged intact at the other end. And Bondarenko seemed to be coping OK too. He had endured ten of his fifteen days and was about to enter the home straight when disaster struck.

Another session of bio-monitoring had just come to an end, and Bondarenko was peeling the sticky sensors from his skin using balls of cotton wool soaked in spirit which he was then lobbing into the waste bin. Except he missed with one of them, and struck the hotplate, where the saucepan was heating. The highly flammable cotton wool ignited, and the oxygen-rich air quickly turned the scene into a conflagration, setting fire to Bondarenko's jumpsuit.

It took the medical team half an hour to depressurise the chamber and get the door open. Bondarenko was prone on the floor, yet somehow still alive. They wrapped him in a blanket and rushed him to hospital, where he was admitted under a false name – Sergeyev Ivanov – and logged simply as an Air Force Lieutenant. The doctor who saw him eventually spoke about that moment publicly in 1984. He described Bondarenko's appallingly severe burns, the fact that the only undamaged area of his skin was the soles of his feet where his boots had protected him, and his own astonishment that a person who had endured so much trauma could still be alive and capable of whispering to the doctor and pleading with him to do something to stop the pain.

Bondarenko died sixteen hours after the fire. He was buried as an Air Force pilot, under a gravestone marked 'With fond memories from your pilot friends', and he was posthumously awarded the Order of the Red Star three months later. We can probably safely assume that his widow and son were looked after by the government as a cosmonaut's family: Bondarenko's wife certainly continued to work at Star City and Alexandre grew up and joined the Air Force.

Nevertheless, despite the fact that Bondarenko appears in the earliest group photographs of the cosmonauts, the story of what happened to him was suppressed for a quarter of a century. The Chamber of Silence was well named. The details of Bondarenko's fate would not be known publicly in Russia until the mid-1980s.

A month before Yuri Gagarin was launched into orbit, the Space Race had claimed its first life, and the risks involved even at the obscure perimeters of this endeavour were now graphically visible, at least to those on the inside of the Soviet programme. It was a moment for chastened reflection.

Not that everybody seems to have felt that way. Dr Vladimir Yazdovsky, then forty-eight, was the Soviet programme's medical scientist and the person responsible for harvesting data from those fortnights in the Chamber of Silence.

'Valentin violated fire safety instructions,' Yazdovsky wrote, coldly, in the wake of Bondarenko's death. 'It cost me a delay in promotion.'

VI. THE FIRE ON THE PAD

Shortly after Mike Mullane, the Shuttle astronaut, joined NASA, he and some of his newly drafted colleagues were taken on a guided tour of the legendary Mission Control Center in Houston. As I mentioned earlier, I was taken around that hallowed room too, as a visiting ESA astronaut, and eagerly drank in the history that is soaked into its screens and furnishings and somehow into the very walls of the place.

But Mullane's tour ended quite differently from mine. Having taken seats at the consoles and put on earpieces and headsets, and having had all the various MCC roles explained to them, Mullane and his colleagues were invited to keep the intercom on for just a little longer. Their guide then cued in a tape. It was a recording of the last moments in the lives of Gus Grissom, Ed White and Roger Chaffee, the crew of Apollo 1, who died in a fire during a training exercise on the launch pad on 27 January 1967.

When the tape finished, everyone in Mullane's group sat in shocked silence for some time.

'The motive of our teacher was clear,' Mullane wrote. 'He was attempting to open our eyes to the reality of our new profession. It could kill us. It had killed in the past and held every potential to do so again.'

In the way that other astronauts felt uneasy about the Lunar Landing Training Vehicle, Gus Grissom had felt uneasy about the

Apollo 1 Command and Service Module. The difference was, this piece of machinery wasn't just intended to fly a few hundred feet above a runway; this one was intended to go to space.

'Glitchy' wasn't the half of it. Grissom maintained that the problems with that spacecraft came 'in bushelfuls'. Both the soft-ware and the hardware needed updating so constantly that the simulator the astronauts were training on couldn't keep up. So continual were his trenchant dismissals of the system and its seemingly endless flaws, the engineers had started referring to Grissom as 'Gruff Gus'.

One morning on his way to work at Cape Canaveral, Grissom picked a large lemon from the tree in his front garden. His wife, Betty, asked what he was going to do with it. He told her he had something very specific in mind. At the conclusion of that day's exercises, he left it perched on top of the Apollo 1 simulator for the engineers and everyone else to see and duly note – an unspar-ing comment about where he stood vis-à-vis the machine that was supposed to take him to space. It was a lemon.

That wasn't the only satirical gesture Grissom had made in the direction of this spacecraft. Along with his fellow Apollo 1 crew members, the Gemini astronaut Ed White and the as yet unflown Roger Chaffee, Grissom had posed for a crew portrait around a small plastic model of the spacecraft, their heads bowed, their eyes closed and their hands together in prayer. They presented it to the Apollo Spacecraft Program Office manager, Joseph Shea, with the inscription: 'It isn't that we don't trust you, Joe, but this time we've decided to go over your head.'

Still, he got on with it. This, after all, was the job, as he knew well. By now – the beginning of 1967 – Grissom had become the first American to fly in space twice, crewing Mercury and Gemini missions. He was many people's tip for the Mercury astronaut most likely to go all the way to the Moon. And now here he was, another step along that journey, commanding Apollo 1, which was to provide a low-orbital test flight for the Apollo Command and Service Module, the first such with a crew on board.

Late in the afternoon of 27 January, that crew assembled for a 'plugs out' test on Launch Pad 34 at Cape Kennedy – a dress rehearsal for the launch, with the rocket unfuelled. Grissom, White and Chaffee got suited up, ascended in the lift and inserted themselves in the cramped confines of the module, sitting 363 feet in the air at the tip of the already wheeled-out Saturn 1B rocket. The hatch was sealed, pure oxygen was pumped into the pressurised capsule and a simulation of the countdown to launch began.

The exercise went badly. The master alarm kept going off, triggered by a high oxygen flow, and interrupting the countdown. At one point, farcically, there was a breakdown in communications with the blockhouse, a matter of yards away. Grissom angrily remarked, 'How are we gonna get to the Moon if we can't talk between two or three buildings?' The countdown was paused, Ground Control were instructed to try to sort out the comms problem, and everyone else was advised to take a short break where they sat.

But then it rapidly got worse. The first sign of the disaster for those monitoring the exercise came on the screen in front of them, showing an image from a remote camera of the capsule's window, seen from outside. The window had been dark. It suddenly flared bright white.

Then a shout came from inside the capsule – the voice of Ed White.

'Fire!'

And then some further shouts.

'We've got a fire in the cockpit!'

'We've got a bad fire . . . get us out. We're burning up.'

And then there was a scream.

Repelled by heat and thick, choking smoke, the pad rescue team only reached the capsule's hatch with extinguishers five minutes later. The interior was already charred. The pure-oxygen atmosphere had fed the blaze. White had fought briefly to open the hatch from inside, but it was impossible. The inward-opening hatch could not be opened at all by human force when the capsule was pressurised. Trapped in that tiny space, Grissom, White and Chaffee died, not from their burns but from asphyxiation.

The fire appeared to have been started by faulty wiring, sparking somewhere under Grissom's feet. The bundles of exposed wire in the capsule had been one of the things that Grissom and his colleagues had complained about, along with the presence of flammable materials generally. And now three astronauts were dead.

As all astronauts must, Gus Grissom had certainly considered the possibility that he might not survive the job – that the job might kill him. Everyone who flies in space needs to make their peace with that prospect to some degree – needs to have some belief in the greater value of the larger cause. Perhaps, given Grissom's scepticism about the safety of his spacecraft, he had considered such things more fully than most. Either way, he had had this to say: 'If we die we want people to accept it. We hope that if anything happens to us it will not delay the programme. The conquest of space is worth the risk of life.'

Those were selfless words, and possibly consoling to some degree in the aftermath of the incident. So too, no doubt, were the lessons learned and the changes made in the year of hard scrutiny and reflection that followed this tragedy, during which the construction of that Command and Service Module was completely readdressed: a redesigned hatch that opened outwards; a 60/40 oxygen/nitrogen mix in the cabin; no more flammable Velcro on flight suits; fireproof beta cloth, not ordinary cloth; high-strength epoxy and Teflon; metal casings around wires; stainless steel plumbing, not aluminium; no more nylon netting or polyurethane, no paper of any kind, anywhere on board.

But the effect of that hugely talented crew's loss, and the circumstances of that loss – on the ground, while training – were still devastating. Flight Director Gene Kranz later wrote: 'We knew there was a high probability that some men would die at some point in the programme, but none of us could accept losing crew on the launch pad. We had all assumed that when calamity struck us, it would be in flight. Our nightmare was an explosion during launch, or a flying coffin, a faulty craft stuck in endless orbit . . . We mourned our crew and the loss of whatever naivety we had left.'

The first American lives sacrificed in the quest to get to space were lost on the ground, during a dry run – a grim reckoning for everyone regarding the height of the stakes, not just in flying to space but even in getting ready to do so.

The Mercury Seven, 1959 press conference: 'Could I ask for a show of hands,' asks a reporter, 'of how many are confident that they will come back from outer space?' Each of the Seven immediately raises a hand, except for astronauts John Glenn and Wally Schirra. Those two raise both their hands.

1961. Tom Wolfe's book about the Mercury astronauts, America's first official draft of space pilots, immortalised the phrase 'the right stuff' to describe the attributes needed to embrace the prospect of blasting out of the Earth's atmosphere for a while and then re-entering it at 24,000mph, with all that journey's inevitable dangers and uncertainties.

Did she cook for her family? Proper meals? How many times a week? Was she a drinker at all? How much? Did she seem to be a good patriot? And did the marriage seem sound? These were just some of the intrusive and outdated questions that the wives of the Mercury astronauts encountered from the press and general public.

Right: US comedian Bob Hope interviewing the Mercury astronauts' wives about space flight training. *Front row, left to right:* Annie Glenn, Louise Brewer, Betty Grissom and Josephine Fraser. *Back row, left to right:* Marge Slayton, Trudy Olson and Rene Carpenter.

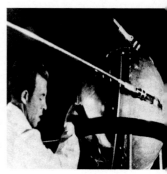

In the United States, getting a human into space had become a matter of national urgency, dramatically, on 4 October 1957, when the communist Soviet Union successfully launched th first orbital satellite, Sputnik 1.

1959. The Mercury Seven posing in their iconic pressure suits.

NASA Mercury astronaut John Glenn. If you are looking for a definition of 1960s astronaut cool, look no further.

Just a month after the Sputnik 1 launc the Soviet Union launched Sputnik 2, this time with a dog on board. Laika – literally 'Barker' – who press releases romantically insisted was a stray from the streets of Moscow.

The decision of who would become the first Russian astronaut came down to such arbitrary and superficial things. It certainly didn't come down to operational competence in the spacecraft: both Gherman Titov (*left*) and Yuri Gagarin (*right*) had that in spades. Rather, there were mild worries about Titov's first name – Gherman. Gagarin, on the other hand, had a fantastic smile, instantly warm and winning. People were constantly remarking on it. He would look great in the photographs.

12 April 1961. Yuri Gagarin became the first human to orbit the Earth.

In the very likely event of him parachuting to the ground in a remote spot, he might easily be mistaken for a downed American spy. So a pot of red paint and a brush were quickly drummed up, and the legend 'CCCP' was applied to Gagarin's white helmet.

Above, left: Just under a month later, as he sat in the capsule of Mercury-Redstone 3 on 5 May 1961 and waited to become the first American in space, the launch pad escape options open to Alan Shepard included an explosive hatch release – a personal parachute which he wore strapped to his back, making the highly congested cockpit feel more congested still.

Above, right: The mission went smoothly, making Shepard a household name in the US. Here he is returning to Earth and being hoisted from the Mercury capsule.

In 1959, NASA explored sending women to space – referred to at the time as the FLATS (First Lady Astronaut Trainees). However, despite the formidable claims of these remarkable women for serious attention, NASA did not officially sanction the programme to be taken any further. The women were later known as 'the Mercury 13', courtesy of a 1990s television documentary, and this photograph was taken for the group's reunion. The programme remains one of NASA's great missed opportunities.

August 1962, Andriyan Nikolayev and Pavel Popovich. Once in orbit, their two spacecraft flew to within 4 miles of each other and established radio contact – the first ship-to-ship communication in space. A tandem launch! American hearts sank.

Another Russian first: Valentina Tereshkova, the first woman in space, 14 June 1963 (on Vostok 5). She had chosen the call-sign 'Chaika' – 'Seagull'. As the violence of the rocket's g forces gripped her, Tereshkova could be heard over the radio repeating that call-sign to herself like a mantra: 'Ya Chaika, ya Chaika' – 'I am Seagull, I am Seagull'.

Ed White on a spacewalk during the Gemini 4 mission, 1965. Occasionally tools still get lost, floating off into space 'on a definite trajectory going somewhere', in the words of Ed White. Spacewalking remains the most physically and mentally demanding task for any astronaut, and the one that carries the greatest risk.

Gus Grissom, Ed White and Roger Chaffee, the crew of Apollo 1, during capsule training. All later tragically died in a fire during a training exercise on the launch pad on 27 January 1967.

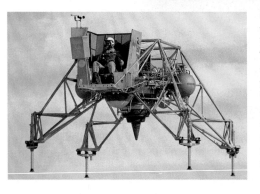

Bill Anders, who flew on Apollo 8, would eventually be able to look back and declare the Lunar Landing Training Vehicle, or LLTV, 'a much under-sung hero of the Apollo programme'. Not, though, before it had almost claimed the life of Neil Armstrong in a dangerous training accident. However, Armstrong's many hours in the Belching Spider would ensure that the Moon landing went smoothly.

Apollo 11 launched on 16 July 1969 and set off on its four-day journey to the Moon.

1969. Neil Armstrong in training for the Apollo 11 mission. Soon, Armstrong would be at the bottom of the ladder and taking his 'one small step for man, one giant leap for mankind' for real – he would become the first human to walk on the Moon.

20 July 1969. This interior view of the Apollo 11 Lunar Module shows astronaut Edwin E. 'Buzz' Aldrin, Jr during the lunar mission. The picture was taken by Neil Armstrong prior to the landing.

20 July 1969. Neil Armstrong's photos of Buzz Aldrin's first steps on the Moon. This was what it all boiled down to in the end, even after the unprecedented feats of rocketry and all the extraordinary technical innovation which had safely flown two astronauts to a destination a quarter of a million miles from home: the basic human act of walking.

Armstrong's, Collins's and Aldrin's reward for enduring just over four days beyond the bounds of Earth was a further twenty-one days of solitary confinement, one of which was Armstrong's thirty-ninth birthday.

Gene Cernan during a later Apollo 17 mission, covered in lunar dust, which has an electrostatic charge and clings to everything. Wearing his white undershirt and with his forehead smudged with sweat and black smears, Cernan looks less like a spaceflight pioneer and more like a coal-miner at the end of a long shift.

CHAPTER THREE

GETTING A MISSION

'If an astronaut had been in space, he was a star. If he was on a crew, he was a prospect. If he was not yet in line, he was simply a suspect. He hadn't really made the team.'

– Walter Cunningham, *The All-American Boys*

I. 'WE ARE *GOING*!'

It's April 2023. In a hangar at Ellington Field, a handful of yards from where Neil Armstrong ejected himself from a stricken Flying Bedstead more than half a century ago, NASA reveals the names of the four astronauts selected as the crew for the Artemis II mission, scheduled to launch in 2024.

The hangar has been transformed for the occasion into a studio – lights, cameras, stage-set, drapes, all ready to pump out a live-streamed news package which will also nod heartily to the production values of TV talent shows: a primed audience, rolling autocues, pre-filmed inserts on giant video backdrops, and stirring orchestral music with synthesised French horns to the fore. It is clear that none of the starring participants today will be lobbed rogue questions about their smoking habits or asked to state their home address by the man from the *Delaware Sentinel*. We are a long way altogether from Dolley Madison House and the announcement of the Mercury Seven.

But then, of course we are. This is the first Moon crew to be assembled in fifty years, the first of the new social media age, a team preparing to go on a ten-day mission across 600,000 miles, pass above the Moon's surface, as the crew of Apollo 8 first did, and return to Earth at thirty times the speed of sound; a crew set to ride the Space Launch System rocket, the most powerful rocket in operation, mated to the Orion capsule, and shake down the systems that will eventually take Artemis III all the way to the surface; the crew who will demonstrate beyond the slightest doubt that crewed lunar flight is back on, and the Moon again within human reach. Some fanfare is clearly due, synthesised French horns and then some.

'Are you ready to meet your team?' asks Joe Acaba, chief of NASA's Astronaut Office and veteran of 306 days in space across three space missions, and the room whoops its affirmation. 'Four names, four explorers,' Acaba continues. 'Four of my friends once more answering the call to rocket away from Earth and chart a course around the Moon.'

And here they come, walking out in their blue flight suits – and also shortly to be seen in individual video clips, this time in their orange Moon mission pressure suits, striding through dry ice towards the camera and folding their arms like sports stars.

Here's Commander Reid Wiseman, who was a flight engineer on the ISS for 165 days in 2014 and is the veteran of two spacewalks. ('Tonto', to use his call-sign, was also the Capcom on the ground monitoring my spacewalk from the ISS in 2016.)* Here's the pilot, Victor Glover, who served a 168-day mission on the ISS in 2020–21, making four spacewalks. Here's Mission Specialist

* To avoid a cacophony of competing voices in the astronauts' ears, Ground Control created the Capsule Communication or Capcom role, handing it to an astronaut who would then be the filter for instructions to the crew and their representative on Earth.

Christina Koch, a true pioneer, who spent 328 days on the ISS in 2019–20, and thereby holds the record for the longest single spaceflight by a woman, beating Peggy Whitson's record of 288 days which had stood since the eighties. Koch also took part with Jessica Meir in the first all-female spacewalk. And here's Mission Specialist Jeremy Hansen, the rookie on the crew, selected, as I was, in 2009, a good friend, and a Canadian: for, as Vanessa Wyche, Director of the NASA Johnson Space Center, explains, 'In the twenty-first century, NASA explores the cosmos with international partners.'

Here, only dimly foreseeable in the Mercury era, is a Moon crew including a woman and a person of colour. 'This is humanity's crew,' NASA says. These are the four people who 'will carry the hopes of millions of people around the world'. This is the crew who will take 'the next step on the journey that gets humanity to Mars' and offer inspiration way beyond the immediate remit of their mission. As the rhetoric builds, we are invited to think of the Moon as 'a symbol of our can-do spirit' and there is a strong and deliberate echo of John F. Kennedy in the assertion of this crew's intention 'to do the things that are hard so that others might follow behind them'.*

NASA Administrator Bill Nelson, the former US Senator who flew as a Payload Specialist on Space Shuttle *Columbia*, says his piece: 'The Artemis II crew represents thousands of people working tirelessly to bring us to the stars. Together we are ushering in a new era of exploration for a new generation of star sailors and dreamers – the Artemis generation.'

---

* 'We choose to go to the Moon. We choose to go to the Moon. We choose to go to the Moon in this decade and do the other things, not because they are easy, but because they are hard. Because that goal will serve to organize and measure the best of our energies and skills, because that challenge is one that we're willing to accept.' JFK, speech at Rice University, 12 September 1962.

And then the room is invited to shout out the Artemis tag-line: 'We are GOING!'

The whole NASA astronaut corps – all forty-one of them – files across the stage at one point, pausing to exchange high fives and hugs with Acaba, before gathering in a huddle just to one side, a crowd scene in blue. The symbolism is clear. Four may have been summoned, but many more comprise the team behind them, lifting them to prominence.

Those unchosen ones seem entirely happy to share the excitement of this big moment for four of their colleagues, and to play their heartfelt part in the whooping, the hollering and the applauding.

But they wouldn't be human, surely, if somewhere in their minds they weren't asking themselves: the next time some of our number get assigned to a crew, what can I do to ensure that it's me?

They wouldn't be human, surely, and they certainly wouldn't be astronauts.

II. FLOWN AND UNFLOWN

Mike Mullane can precisely date the moment at which the mood in his astronaut group – the 1978 draft, known familiarly as 'the TFNG', the 'Thirty-Five New Guys' – changed entirely. It was the day George Abbey, who was then the formidable chief of NASA's Flight Crew Operations Directorate, called everyone together and announced the group's first mission assignments.

On the crew lists for the next three slated Space Shuttle flights were seven of the thirty-five.

'Hopefully, we'll get more people assigned soon,' Abbey said, after he had read out the names. And then he left the room.

'*Poof*,' Mullane wrote. 'With Abbey's words, TFNG camaraderie vaporised.'

This was 1982. For more than three years, Mullane and his thirty-four peers had been training together, learning together, flying jets together, gathering together at the Outpost Tavern in Houston for beers at the end of the working day – energetically living the astronaut life. And for that whole time they had been united by a common bond: the fact that all of them were as yet unassigned to a mission in space.

But not now, not since this announcement by Abbey, their revered and feared overlord. Now seven of them had been handed the golden ticket that all thirty-five of them had longed for. Seven of them suddenly had their names on the manifest for a flight. And, of course, by extension, twenty-eight of them did not. Twenty-eight, including Mike Mullane.

From that point on, Mullane claims, the group was unavoidably two groups – the assigned and the unassigned; the soon-to-fly and the still-uncertain; the forging-ahead and the left-behind.

'I tried my best to be rational,' Mullane wrote. 'Somebody had to be first. We couldn't all be. But I couldn't accept the rationale and I doubted any of the others could either. We were too competitive.'

Mullane says that the thirty-five never again gathered socially, all together, after that day. The split proved irrevocable.

It's practically impossible to exaggerate how much the matter of securing a mission dominates the day-to-day thinking of the trainee astronaut. There's an awful lot of other stuff to be thinking about, it's true, and much of it extremely absorbing. Yet for as long as an astronaut remains without a confirmed seat on a confirmed rocket departure, questions will be constantly flickering around the periphery of his or her mind: questions about what needs to be done, and who needs to be convinced, and how and when, to bring about the desired outcome of a categorically fixed assignment.

Because the basic fact is, there is nothing in your contract as an astronaut that guarantees you will fly in space. It's easy to overlook this in the excitement that surrounds getting through the interview process and being selected for the corps. And, yes, that ultimate outcome is partly up to you and how well you complete your training; it is in your hands to that extent. But – and here's where the frustration and the anxiety, even the paranoia, can creep in – it's never going to be entirely about that. It's also going to be about forces and outcomes and decisions and contingencies that you can't control and that you possibly don't even quite understand.

Meanwhile, right behind these concerns, and directly fuelling them, hovers the haunting figure of the unflown astronaut. Every space agency has one: the trainee who seems to have been knocking around for quite a while and doing everything required of them to stay in the job . . . yet has never been to space.

Inevitably, there is a certain poignancy attached to that predicament, the sense of a thwarted desire, right out there in the open, and indeed coming into work every day. It could almost be the stuff of mournful folk songs – 'The Ballad of the Unflown Astronaut', or something. No astronaut can contemplate that predicament without experiencing a shiver of dread.

And every trainee astronaut knows that, until the moment comes and a rocket launches with them inside it, and until that rocket reaches the Karman Line 100km above the Earth – the point at which space is officially agreed to begin – they too are an unflown astronaut, and their story is incomplete.

'Remember – it's better to have a spaceflight in front of you than behind you,' Michel Tognini, who was then head of the ESA Astronaut Office, told us all shortly after we joined the agency. And all six of us sat there and nodded politely while, inside, disagreeing violently. I'm sure we all still do. I appreciated the sentiment, but better to be a flown astronaut than an unflown one, every time.

I came to know the fear of not flying – that shiver of dread – very intimately during my own pre-flown days. At the European Space Agency there is, relatively speaking, an openness and clarity about the assignment system which is very different from the way they order these things at NASA. ESA is funded by individual European governments, and contributing nations essentially purchase missions for their astronauts with the strength of their financial commitment. The more a country stumps up, the more it can demand by way of pay-back in terms of flights for its astronauts. Consequently, even ahead of the assignments being formally declared, it was always relatively easy for me to look ahead and work out the likely running order and assess my own position on the runway in relation to my five ESA colleagues in the draft of 2009.

And that position, I should further say, was fairly obviously right at the back of the line.

At the point at which I joined, the UK was not a contributor to ESA's Human Spaceflight programme; it paid into other ESA programmes, but not that one. Accordingly, it was really the work of seconds for me to calculate that I was last in the queue of the astronauts in my draft. I would have known this even if I hadn't gone to use the printer in the ESA offices one day and found an official management document left in its tray which explicitly stated as much.

Well, things get left in printer trays – it happens. But that was a little careless of someone, and quite the thing for me to read in black and white, my name leaping off the page at me.

'For the moment it is most probable that Timothy Peake from the UK is the one with the lowest chance to fly . . .'

But ESA's problem at that point was a simple one. They had

secured three missions to the International Space Station through NASA, along with two further missions separately bartered for by the Italian government, and clearly destined to be flown by the two Italian astronauts in our team, making five missions for ESA astronauts in all between 2010 and 2019. But they had drafted six astronauts, meaning they had one too many – the one in question being me. At that moment I was officially 'the reserve astronaut', which wasn't quite like being 'the unflown astronaut' but felt horribly close to it.

And I was likely to remain that way for nine years, at the end of which there was still no guarantee that I would fly.

I realised I needed to adapt my feelings about the status of the unflown astronaut, because, the way things were going, that looked very likely to be me. I had to think positively about what I was being shown as a trainee in the meantime: the solar physics, the chemistry and biology, the space law, computer technology, electronics, not to mention the incredible places the training journey was taking me, some of which I mentioned in the previous chapter – deep into Sardinian caves, out to Houston for spacewalk training, down into the Aquarius research station 25 metres below the sea, over to Russia, out to Japan . . .

Basically, I decided to embrace the training experience in itself, in case that was all this particular voyage amounted to. If it had been me writing 'The Ballad of the Unflown Astronaut' at that point, there would have been a line in there somewhere about how even an unflown astronaut was something to be. And I would have meant it.

Another useful thing about ESA's way of doing things: if I hadn't flown, I would have had the consolation of being able to think that it wasn't about me. It was no reflection on my ability or otherwise as an astronaut. I could simply put it down to those tiresome old foes, politics and funding. What are you going to do?

But at the same time, for all my diligent work on constructing and maintaining a brave face, I know that if things had worked out differently – if the UK government hadn't eventually come on board and increased its funding, and if a long-duration mission to the ISS hadn't come my way in 2015 – it would have gnawed at me for the rest of my life.

Still, all of these stresses are very much easier to cope with if the parameters surrounding your selection seem clear to you – if you know what you're up against and what stands in your way. In the eighties-era NASA system that Mike Mullane was part of, that wouldn't have been the case. There, nobody's individual fate was preordained or made swiftly discernible by anything as clear-cut as funding issues. Everybody on the corps was in an open and long-term competition for assignments.

I'm not sure how much I would have enjoyed being part of an astronaut office that was set up to be so fiercely competitive. Whatever you wanted to say about the ESA system, it drew the sting from the competition between us individually and snuffed out the second-guessing at source. We all knew the score, and a social fracture in the team such as the one Mullane experienced never occurred in mine, even as the assignments came and went.

But it wasn't just competition in itself that introduced distress into that NASA astronaut class. Competition can be healthy, after all. It was the fact that the rules of the particular competition everyone found themselves engaged in never quite seemed clear. Mullane wrote about his sense of rejection after those first names were called, but also about his bewilderment: 'I had to believe I was at the end of the flight assignment line, and, most maddening, I had no idea how I had gotten there or how I might recover.' And that bewilderment seems to have arisen as the consequence of a system that placed Mullane's fate as an astronaut exclusively in the hands of one extraordinary power-broker.

Historically, mission assignments at NASA had been in the gift of some formidable individuals. During the Mercury phase, it was Bob Gilruth who as General Mission Commander made the decisions on who flew and when. By the time of the Apollo project, that power had fallen to Deke Slayton from the Mercury Seven.

Slayton's chances of getting to space on a Mercury mission came to an abrupt end one day in August 1962. During a routine session in the centrifuge in Johnsville, the medics spotted something on his heart monitor which they thought needed investigating. Slayton was sent for further tests. He was meant to be with the other members of the Mercury Seven on a factory trip to Convair in San Diego at that point. NASA, not wanting the word to get out, covered for him by announcing that he had a viral infection. The tests found him to have mild idiopathic atrial fibrillation – an irregular heartbeat. Just like that, Slayton's career as a Mercury astronaut was over.

One can only imagine how shattering this blow must have been for him, its timing especially cruel. Arguably, after John Glenn, Slayton seemed the member of the Seven most built for a certain kind of astronaut stardom. He had the Hollywood looks and the understated, slightly detached manner.

Also, Deke Slayton? Even his name seemed to come from central casting.

The schedule strongly implied that Slayton was going to fly on Mercury-Atlas 7, following Glenn in America's second crewed orbital mission. Had he managed it successfully it would have guaranteed him hero status, ticker-tape parades and lasting financial security. But now he was benched.*

And what for? Many of the doctors Slayton now consulted in desperation for a second opinion seemed to think his condition

---

* The MA-7 mission went to Scott Carpenter.

was trifling, that there was nothing about it that compromised him as a pilot, nor as anything else. But the view of NASA's medics was unsentimental: if six other astronauts were available who didn't have an irregular heartbeat, why would they risk flying one who did? Slayton, to his bitter and lasting disappointment, was stood down, and the Seven became Six.

His consolation prize was being asked to run the astronaut office and be 'coordinator of astronaut activities'. His first task was selecting the next intake of astronauts, the group of nine that included Neil Armstrong, Frank Borman, Pete Conrad and John Young. Slayton would then come to assume almost absolute power over the selection of crews – who got to fly and who didn't, and what part they got to play in the mission.

He also got to fly himself one day – but that's for later in this book.

According to Gene Cernan, 'With few exceptions, Deke always made the final call on who flew, when, and in which seat, and we studied his choices with great care, looking for some pattern. But there wasn't one.' This was the exasperating thing. The astronauts would tie themselves in knots trying to second-guess Slayton and looking for the clues in his choices that might throw some light on the best way to go about finding his favour and getting a mission. But if there was a code, nobody ever cracked it. It was just down to Deke.

By the time Mullane's career was underway, in the mid-1980s, the power over assignments had passed to George Abbey. An electrical engineer and US Air Force pilot, Abbey applied to be an astronaut in 1964 but was rejected. However, he took an engineering job with NASA and thereby began a remarkable achievement-studded career with the agency that would run for thirty-nine years. He was part of the team that redesigned the systems after the deaths in the Apollo 1 fire and he was also a member of the ground team which dealt with the Apollo 13 emergency. On the back of that work, Abbey received the Presidential Medal of Honor from President Nixon in 1970, when he was just thirty-eight.

The ground-breaking 1978 astronaut draft, the Thirty-Five New Guys, which properly opened the doors for women and civilians, was made under Abbey's supervision. It put the first American woman in space (Sally Ride), the first African American in space (Guion Bluford) and the first Asian American in space (El Onizuka), and it produced the first American woman spacewalker (Kathy Sullivan). In any list of the most significant drivers of culture change at NASA, Abbey would be somewhere around the top.

Yet he seems to have been a hard figure to pin down – quietly spoken, to the point where people had to crane forward to catch his words, evasive, disinclined to meet anyone's eye. He was a strong presence yet oddly absent at the same time. Despite his prominence in the organisation, he could very easily pass under the radar beyond the perimeters of the Johnson Space Center.

I met him just once, in the legendary Chelsea Wine Bar, a regular astronaut hang-out in Houston (extensive wine and beer list, cheese and meat boards, deck, water-views), where the quietness of his voice had me and the two ESA rookies I was with leaning forward practically out of our seats to catch his words of wisdom. At least I think they were words of wisdom. On reflection, he could just have been ordering another drink.

Abbey was made Director of Flight Crew Operations in 1983, a role in which he became directly responsible for the selection of Space Shuttle crews until 1987. Among those with direct experience of him during that period, few seem to have had unmixed feelings. Most of the reservations seem to arise from his leadership style, which was hard to read, to say the least. Always dressed for work in a jacket and tie, which immediately set him apart from the flight suit-wearing community, Abbey occupied a corner office on the eighth floor of the NASA administration block at JSC, with a clear view of Building Four, where the astronauts trained. It was felt that he looked down from that remote but lofty position, watching the trainees' every move – a deity, but ultimately mysterious, as deities so often seem to be.

Astronauts referred to him as 'King George' or, less favourably, as 'Darth Vader'. Though he was clearly making the key decisions in relation to their professional lives, he was not inclined to explain himself if he could avoid doing so. The fact that Abbey, rather than John Young, who was Chief of Astronauts, held all the strings when it came to Shuttle assignments puzzled many of the trainees, who felt that Young, a bona fide spaceflight legend, knew and understood them better than Abbey. Some members of the corps decided the Thirty-Five New Guys' group acronym, TFNG, actually stood for 'Thanks For Nothing, George'.

In 1981, Alan Bean, who walked on the Moon in the Apollo 12 mission in November 1969, had to go to Abbey's office and tell him that he had decided to leave NASA after his long and distinguished career, and concentrate on being an artist instead.

'If he hadn't had the window behind him, he would have gone over backwards,' Bean recalled.

Abbey recovered his composure enough to ask quietly, 'Can you earn a living at that?'

Bean replied that, if things got difficult, he supposed he could always try and get a job in a hamburger joint.

The conversation doesn't seem to have gone much further.*

---

* Bean's paintings of the Moon would sell in his lifetime for tens of thousands of dollars. A painting he completed in 2000, in textured acrylic with Moon dust on aircraft plywood and titled 'Jack Schmitt Skis the Sculptured Hills' – a

Abbey's withdrawn manner, his evasions and absolute refusal to open up about his rationale at any point seem to have been the cause of much unease and even anguish in the corps, or certainly for Mike Mullane, who described it as 'a cancer on astronaut morale'. According to Mullane, a third of the office skipped Abbey's leaving party.

Yet he was the person who had at his disposal your chances of escape from the dreaded status of unflown astronaut, so inevitably many trainees embarked on attempts, variously doomed, to catch his eye and curry favour. And perhaps none more spectacularly than astronaut candidate Jim Bagian.

On Abbey's birthday one year, Bagian dressed himself in a Superman outfit, climbed a rope and appeared outside Abbey's eighth-floor window, where he sang a chorus of 'Happy Birthday' before shinning back down again. This sing-o-gram stunt somewhat backfired when the Space Center's unamused security staff complained angrily about it, initiating an in-depth investigation. Nevertheless, Bagian went on to fly two missions on the Shuttle, so maybe it worked at some level.

What definitely needs to be acknowledged is that the Space Shuttle era ushered in a boom time for astronaut recruitment at NASA, and by extension an unprecedentedly complex period for the administration of those astronauts. The Shuttle's 135 missions created 852 individual assignments which went to 355 crew members – 306 men and 49 women – and at its peak, in 2000, the astronaut corps would grow to be 149 strong. Imagine *that* little lot trooping across the stage to high-five Joe Acaba at a crew announcement.

---

response to Schmitt's remark on the lunar surface 'Too bad I don't have my skis' – recently sold posthumously for $368,000. Of his career shift, Bean once said, 'I think everyone can do more than one thing in his life.'

And for all the evasion, and the mumbled conversations, and the feelings of frustration and powerlessness, every astronaut George Abbey was responsible for bringing into the NASA corps in that period was assigned to at least one mission in their working lifetime. None of his trainee astronauts ever went unflown.

But, of course, Mike Mullane wasn't to know that at the point when Abbey called out that first batch of names from the Thirty-Five. Even then, though, in the thick of his own despondency, Mullane had the wherewithal to look around the room and notice colleagues on whom this news had fallen even harder than it had on him.

Fred Gregory in particular seemed to be struggling with it. But that, Mullane realised, must have been because Guy Bluford's name was on the list for STS-8 – which meant that Gregory's quietly nurtured dream of being history's first black astronaut had just been extinguished.

And what about Judith Resnik? It must have occurred to her that she had a shot at being the first American woman in space. With the selection by Abbey of Sally Ride for STS-7, that door had just closed on Resnik and the four other women in the 1978 draft – Anna Fisher, Shannon Lucid, Kathryn Sullivan and Rhea Seddon. Which also meant the door had closed on a prominent place in the history books and everything that might follow from that for the rest of their lives.

Sometimes there was so much more at stake in selection decisions than just a flight to space.

III. DEATH OR GLORY

It was early April 1961, and the decision was keeping Nikolai Kamanin awake at night. 'All the time I am haunted by one thought,' the head of Soviet cosmonaut training wrote in his diary. 'Who should I send, Gagarin or Titov? . . . It is difficult to decide which one of them to send to their certain death, and it is just as difficult to decide which of these worthy men should be made world famous, whose name will be forever preserved in the history of mankind.'

After more than a year of solid training, it had become obvious that the choice for the first cosmonaut in space was between Yuri Gagarin and Gherman Titov. Along with the other four members of the Vanguard Six, they had spent January doing simulator tests and going through an exhausting bout of parachute and recovery training exercises. That had been followed by one final set of examinations, assessing their fitness and readiness to fly, evaluated by an inter-departmental committee. And Gagarin and Titov had come out on top.

But which was it to be: the smiling Gagarin or the more serious-looking Titov?

And which was it to be: death or glory?

When Kamanin wrote about sending one of these men to their certain death, he articulated fears which would naturally arise from what he (in common with everyone else) didn't know about space – this place where no one had yet been – and about launching people there on top of enormous rockets.

But then there was the additional factor of what he *did* know about the Soviet spacecraft.

On both sides, the urgency of the Space Race led, inevitably, to shortcuts and oversights and a certain amount of sweeping of problems under the carpet. But it would be fair to say that it especially led to those things on the Soviet side.

For example, in sea trials, the emergency dinghy stowed on board the Vostok capsule had revealed itself to be pretty much hopeless, unstable and prone to leaks. There was little doubt that it would sink if called upon to do duty in anything more choppy than a garden pond – and possibly even there, too.

But was the programme halted while engineers sorted that problem out? No. The Soviet agency bashed on. I mean, the chances were the Vostok dinghy wouldn't be needed anyway, right? (Indeed, it wouldn't.) On the Soviet side there was a disinclination to sweat the small stuff and a less (shall we say?) fussy attitude to operator safety than displayed by the Americans, who were by no means unhurried themselves.

Actually, there was disinclination to sweat some of the bigger stuff, too. Incredibly, a manual override for re-entry controls – enabling the cosmonaut to orient the craft's angle of re-entry and to activate the braking engine in the event of a computer malfunction – was only added to Vostok at the last minute. When the six cosmonauts arrived in Baikonur prior to the scheduled launch, they had just one week to practise manual re-entry, something the Mercury astronauts had been training to do for two years.

Furthermore, the Vostok's manual override system came with unusual levels of confidentiality built into it. It was to be triggered by inputting a secret three-number code. The cosmonaut was to have no prior knowledge of that code, just in case he went rogue and took control of the spacecraft for his own undesirable ends, such as defecting and thereby humiliating the USSR.

In a detail which has something of the flavour of a budget gameshow about it, the all-important code numbers were to be concealed on a piece of card in a sealed envelope, which was then going to be glued into the cockpit's inner lining, within reach of the cosmonaut's seat. In the event of an emergency, the cosmonaut could then be alerted to the code's whereabouts and could rip into the lining, tear open the envelope, input the code and take control of the spacecraft.

Did this seem a bit . . . fiddly? Would it seem so in an emergency, playing hide and seek for a potentially life-saving envelope while trying to regain control of a spacecraft? Nobody seemed to mind.

However, Vostok would not, at least, be flying with an 'Emergency Object Destruction Device' on board – 'Emergency Object Destruction Device' being a splendidly euphemistic term for what you or I might refer to in less formal circumstances as a bomb.

A bomb for emergencies had been a feature of the uncrewed Soviet flights, and the flights that contained dogs. In the event of everything getting out of control and the spacecraft falling into the wrong part of the world, Ground Control could simply press a button and unsentimentally blow the stray vehicle up, thereby preventing it from causing unwanted destruction but also ensuring that the dog and/or sundry other critical contents did not fall into the wrong hands.

Better a purposefully destroyed craft – even a purposefully destroyed dog – than a loss of competitive advantage, went the thinking here.

Apparently the deputy chairman of the KGB, Pyotr Ivashutin, wanted the bomb option to remain on the crewed flights too. After all, he reasoned, there were now going to be some manual controls on the spacecraft – albeit operated by a secret code sealed in an envelope and stuck behind the lining of the cockpit. Nevertheless, what if some disaffected cosmonaut cunningly managed to acquire the code, push the override buttons at the right time and set sail for the West, never to be seen again? Would it not be wise, Ivashutin seems to have argued, for Ground Control to retain the capacity to nip such behaviour in the bud by blowing any such rogue cosmonaut out of the sky?

It was an idea, clearly. However, Ivashutin did not prevail. Indeed, he seems to have been the only person around the table who was completely at peace with the idea of eliminating a Russian cosmonaut in a remote-controlled explosion. It was decided to leave the EODD off the spacecraft.

One other area for concern: what if things went wrong on the launch pad? How could the cosmonaut escape a terrible and agonising death if the rocket didn't launch but simply caught fire?

Well, the Vostok capsule had an ejection system, which would fling the cosmonaut out of the capsule. And then there was a steel net, erected over part of the flame pit – the bit where, more or less, given a fair wind, a cosmonaut ejected from the capsule might come down.

But of course, as mentioned before, that Baikonur flame pit was a concrete canyon 45 metres deep – so what then? How was the cosmonaut to get off the net?

Well, there was a domestic bathtub, attached to a rope. The ejected cosmonaut would crawl across the net and climb into the bathtub. Then, in due course, and when it was convenient, a rescue team down below would use the rope to lower the bathtub to the ground.

Was this evacuation system ever tested? Apparently not, or certainly not by any cosmonauts. In fact none of the cosmonauts had had even the *existence* of this evacuation system revealed to them during training. Assuming he survived the backwards ejection, and assuming he landed on the metal net, and not somewhere where the metal net wasn't, and assuming that landing on the metal net didn't in itself kill him, the cosmonaut was just going to have to improvise – spot the bathtub, maybe, and think to himself: 'Ah, there's a bathtub hanging there, attached to a rope. Maybe I could climb into it and it could become the vehicle by which I am lowered to the ground.'

The author Stephen Walker, in his superlative account of Gagarin's first flight *Beyond*, records that Titov was told about the existence of this escape system many years later, at a dinner, and that he laughed so much he wept.

And then he stopped laughing and weeping and said, very earnestly, 'It's good that we did not know this.'

Anyway, against this backdrop, Kamanin now had a particularly agonising choice to make. It didn't exactly help that the two cosmonauts were close friends, and had been since the start of training. Their families lived next door to each other. When Titov and Tamara, his wife, lost their son, Igor, at eight months, it was Gagarin and his wife Valentina who comforted them and tried to help them through it.

In many ways, the first flight was the lesser deal. The first flight was to be a single orbit. The second flight was intended to kick on from there and rack up sixteen orbits. It was clear which was the greater and more demanding mission. Yet, of course, going first had its own precious kudos which could outweigh all other considerations. History and life-changing quantities of fame and attention were up for grabs in this single, simple selection decision.

Titov was, in Kamanin's opinion, marginally the better cosmonaut than Gagarin; but that only made Titov a better candidate for the second flight – the longer mission. Gagarin, on the other hand, had a fantastic smile, instantly warm and winning. People were constantly remarking on it. He would look great in the photographs . . .

The decision came down to such arbitrary and superficial things. It certainly didn't come down to operational competence in the spacecraft: both Titov and Gagarin had that in spades. Rather there were mild worries about Titov's first name – Gherman. Was that a bit too close to 'German'? And with the Second World War still fresh in minds, wouldn't that sound a slightly off-key note in the great national fanfare that was envisaged when (and if) this mission came off?

Yuri: now, that was more like it. Properly Russian.*

Plus Gagarin was the son of a carpenter – solid Russian stock. He had been a worker in an industrial foundry before he joined the Air Force – again, an unfussy labouring/military background. Titov's father, by contrast, was a teacher. Nothing wrong with teachers, of course, but that's white-collar work. Also, Titov himself seemed to have acquired some slightly learned airs. He was, for instance, given to quoting Pushkin, which most of his fellow cosmonauts probably were not. He could seem a touch distant, too. When the cosmonauts went off together to socialise – going to see a movie, for instance, or heading off to the lake to swim – Titov, perfectly content with his own company, would often politely decline to join them.

The two men were summoned to Kamanin's office in Baikonur, and the decision was announced to them. Gagarin would fly first, and Titov would be his back-up.

Before leaving Star City for Kazakhstan, Gagarin had told his wife that he thought he might be the one selected to fly. She asked him anxiously when the launch might be. He told her it was set for 14 April, deliberately adding a couple of days to the officially scheduled date so that she could be oblivious and not worry about him on the actual day. It had proved to be a wise precaution.

Titov was devastated. In one of the last interviews he gave before his death in September 2000 he said: 'Yura turned out to be the man everyone loved. Me, they couldn't love. I'm not loveable.'

You could understand his bewilderment. He had thought it

* Of the six in the Vanguard, Pavel Popovich must surely have worked out very quickly that his chances of becoming the first person in space were all but non-existent. He was an excellent cosmonaut, and apparently popular and admired within the corps – a singer of folk songs and an excellent wrestler. But he was Ukrainian. That historic first seat was always going to a Russian. Popovich ended up flying fourth.

would come down to training, competence, ability. But those things turned out not to be so rare, or not around Star City. Instead, the decisive quality appeared to be . . . *loveability*?

Could that be right?

IV. TEAM RED V. TEAM GOLD

At the end of 1960, Bob Gilruth asked each member of the Mercury Seven privately to rate their six colleagues and rank them in order of qualification to be the first American in space. Gilruth wasn't, he explained, entirely handing over the selection to the astronauts, just polling opinion to get a broader view . . . and perhaps to see if it matched his own.

John Glenn wasn't very happy about this 'peer vote' idea. He thought he had made all the right moves to clinch that prized first flight – as long as the decision was taken by the people in charge. The support of his peers in a popularity contest? That he felt less confident about, as someone who had spent the last two years making all the right moves might well do.

For his part, Deke Slayton acidly remarked: 'It wasn't the last time John ran for something without having the votes he needed.'*

Anyway, reluctantly or otherwise, everybody handed in their secret ballots to Gilruth. And then, on Thursday 19 January 1961 – which happened to be the day before the inauguration of President Kennedy – Gilruth summoned the Seven to his office and announced his decision.

'This was the most difficult choice I've ever had to make,' he

---

* Slayton was referring here to Glenn's later political career, in particular, presumably, his failure to win the nomination as Democrat candidate for Ohio in 1970. He won it in 1974, though, and was in the Senate for two decades after that.

told them, and added that it was based on three things: evaluations from training officers, the recent peer review that they had all taken part in, and Gilruth's own opinion. How the weighting was adjusted between those three estimates Gilruth never explained, but suffice it to say that he had the last word.

Al Shepard would fly first, Gus Grissom second, and John Glenn third.

The office was apparently silent for a few moments while everyone absorbed this news. Shepard very carefully stared expressionlessly at the floor. Then, one by one, the other six shook the victor by the hand and left.

After standing on his own for a while, Shepard left too, and drove home to his wife, Louise. When he walked into the sitting room, she already knew what he was going to say from the look on his face, which on the drive home had grown some way from expressionless.

'You got the first flight,' she said.

He hugged her.

'Lady,' he told her, 'you can't tell anyone, but you have your arms around the man who'll be the first in space.'

Louise replied dryly (and also presciently), 'Who let a Russian in here?'

NASA were in no hurry to tell the world this news. As well as immediately placing Shepard and his family at the eye of a storm of media interest that could only be a hindrance as the flight date approached, a bold announcement in advance would complicate NASA's freedom to change things at the last minute if the necessity arose. Yet the world needed to be told something, so NASA put out a press release announcing that those three astronauts, carefully named in alphabetical order – Glenn, Grissom, Shepard – had been selected as the 'prime candidates' for the first flight, with a further announcement to follow about who was the primest of them all.

According to Gene Kranz, Assistant Flight Director for Project Mercury, the decision to select Shepard not only disappointed but even confounded the other six. They were, Kranz says, 'fiercely competitive. Each one of them was determined to be the first man in space; each believed his performance during the months of training and testing would win him the coveted prize.'

But perhaps the one most significantly jolted by the choice was John Glenn. Right from his seamlessly charming and effortlessly diplomatic performance at the original press launch, Glenn had reason to believe that he had positioned himself practically unassailably to be the first to fly. The press seemed to agree with him. Having singled him out behind the table at Dolley Madison House as the first among equals and the leader among leaders, they remained loyal to their hunch thereafter. Now the word was out that a decision had been taken, the press simply filled in the gap: it was going to be Glenn.*

And didn't the alphabetical order on the press release actually say as much?

As he went about the public aspects of his business during this period, Glenn now often found himself being warmly congratulated on his selection to be the first person in space, and asked to anticipate exactly how thrilling and life-altering that moment was going to be for him. This must have rankled with him, given that he knew he was presently running third and was officially unable to correct anybody's impression. According to Slayton, Glenn now embarked upon a personal lobbying campaign, throughout NASA and beyond, to get the decision changed,

* Gene Kranz offers the theory that the press kept on about Glenn not because of their personal preference but as a tactic, figuring that if they hammered away at the wrong name for long enough they would eventually force NASA to correct them. This is plausible, too.

until seniors at NASA quietly suggested he calm down and get back in line.

The press thereafter, following the lead of *Life* magazine, declared that there were now two teams in the Mercury Seven: Team Gold, who were the anointed three, and the rest, Team Red – Schirra, Carpenter, Cooper and Slayton. They also began to infer a Mullane-style split in the group down those lines, which as inferences go from a distance, and given the competitiveness of astronauts, was not implausible. These stories of internecine strife took off to such an extent that the four members of Team Red were instructed by NASA's public relations office to give a press conference with the purpose of informing the world that they weren't unhappy, that there was no Team Red/Team Gold division, and that everyone was busily and contentedly getting on with their work.

Slayton later insisted that the four of them *were* genuinely unhappy – but only about having to give a press conference to state that they weren't unhappy. It's impossible, however, to believe that no egos at all got bruised in the making of that historic decision, given what we know about selection processes, and what we know about astronauts.

Somehow the truth about Shepard's primeness had still not emerged when 2 May and the scheduled launch came round. The Soviet Union had crushed American egos a handful of weeks earlier by putting Yuri Gagarin into space – and not just into space but into *orbit* – but the fervour for discovering the identity of America's chosen one remained heated. Members of the press swarmed the Holiday Inn at Cocoa Beach where the launch team were staying in the hope of being given a tip-off or spotting something revealing. Yet the secret held. The photographers gathered in genuine ignorance at the door of Hangar S soon after dawn on the Tuesday morning of the launch, waiting to see who would emerge in the silver suit and climb into the van that was waiting to take them down to the launch pad.

But the weather intervened – too wet and windy. The launch was scrubbed and put back two days, to 4 May. (It would be scrubbed again then, too, and postponed a further twenty-four hours, to 5 May.) Still hidden in Hangar S, Shepard was helped out of his suit and handed a glass of brandy with which to wind down. At that moment, during the post-scrub toings and froings, a reporter near the hangar door caught sight of him, and at last the story was out.

The headline went out all over America – not 'Astronaut Caught Knocking Back Spirits Before Breakfast Time' but 'First US Attempt to Put Man In Space Postponed: Shepard Given First Call'.

As with Gherman Titov in Russia, to John Glenn would go, in many ways, the greater gift of a longer, more complex orbital mission, and not all that much later. And Glenn would earn his own immutable standing in the history of spaceflight. But at that point he knew that Shepard was about to lay his hands on something Glenn now never would: the satisfaction of claiming a prize that could never be taken from him or surpassed – that of being the first of their class to fly.

And would Glenn have swapped places with him, there and then, even knowing what the future held for him?

Of course he would. He was an astronaut.

V. RIGHT STUFF, WRONG SEX

Sometime in the middle of 1961, Nikolai Kamanin heard a rumour that the Americans would shortly be in a position to assign a woman to a space mission. This news needled him. The head of cosmonaut training wrote in his diary: 'We cannot allow that the first woman in space will be American. This would be an insult to the patriotic feelings of Soviet women.'

So, in order to spare the pride of Soviet women everywhere, a national search for female cosmonauts began. In January 1962, the names of 400 prospective civilian women candidates were gathered by the All-Union Voluntary Society for Assistance to the Army, Air Force and Navy and put forward to the space agency, and within a month Kamanin had whittled that group down to five. Their names were Tatyana Kuznetsova, Zhanna Yorkina, Irina Solovyova, Valentina Ponomaryova and Valentina Tereshkova, and the Soviet Union now had a female cosmonaut corps to train alongside the men in Star City.

Were the Americans *actually* getting ready to assign a woman though? It seems that Kamanin had got wind of an experimental project being run by Dr William 'Randy' Lovelace, whom we met earlier, the man responsible for the medical screening of the Mercury candidates. Lovelace had become curious about how women would cope with the kind of physical and mental testing his clinic had recently been putting men through. It was his hunch that they would cope equally well, but he was determined to find out.

Now, this could be construed, from the distance of the present, as a progressive piece of thinking on Lovelace's part. However, some contextualisation might modify that view a little. These might have been the earliest days of spaceflight, with the very first pioneers travelling solo into the skies, but Lovelace was already envisaging a future in which the world would assemble orbiting space stations and sustain a constant human presence in space. He was right about that, of course. But Lovelace figured that these stations would be huge, and populous, and would require vast numbers of staff doing all sorts of jobs to keep them up and running. And many of those jobs, Lovelace's thinking seems to have been, would surely be the kinds of jobs that women did.

It's something of a moot point, then: was Lovelace interested in exploring female suitability for spaceflight because he saw no reason why women shouldn't challenge men for world-changing roles at the tips of rockets as space pioneers? Or was it because he thought that, in the future, space was going to need secretaries and nurses?

Maybe we should leave that question hanging. The fact is, whether or not he casually shared stereotypical 1950s views of the kinds of jobs women might do, Lovelace was a reputed and influential medical authority who was prepared to take women's physical capabilities seriously at a time when reputed and influential medical authorities tended not to. To that extent he surely deserves to be remembered as boldly forward-looking. Anyway, he now asked a pilot called Geraldyn 'Jerrie' Cobb if she would be interested in coming to New Mexico and assisting him with his enquiries.

Cobb was already an aviation legend. Said to have flown her first aircraft – her father's 1930s open-cockpit biplane – at the tender age of twelve, and to have done leaflet-drops on behalf of travelling circuses from a Piper J-3 Cub at the not much less tender age of sixteen, Cobb had qualified for a pilot's licence by seventeen, and for a commercial pilot's licence by eighteen. During the 1950s, while in her twenties, she set world records for non-stop long-distance flying and for speed and altitude in light-weight aircraft, flying at 37,010 feet on one occasion. At the point when Lovelace got in touch with her, she had experience of flying sixty-four different types of propeller-driven aircraft. By any definition of the right stuff, Cobb appeared to have it – with the wrinkle, of course, that she had no military test pilot experience, NASA's sine qua non for astronaut qualification.

Clearly feeling that there was a point to be proved here, and that she was the woman to prove it, Cobb now went to Albuquerque and, under Lovelace's supervision, completed all three stages of the Mercury evaluation process, physical and psychological. Probably to the enormous surprise of all except people who actually knew her, she sailed through. Indeed, her numbers ranked her in the top 2 per cent of the pilots submitted for Mercury selection.

With Cobb's help, and with financial backing from Jacqueline Cochran, another record-breaking racing pilot and a successful cosmetics businesswoman, Lovelace decided to expand his experiment. Twenty-four other women came forward and underwent the first phase of the Mercury evaluation process, and twelve of those passed and declared themselves willing to go on and attempt the second and third phases, as Cobb had done.

Among them was Myrtle Cagle from North Carolina. Cagle had held a pilot's licence since the age of fourteen. Then there were the twin sisters Janet and Marion Dietrich, who flew air races together. There was Jean Hixson, who in 1957 became the second woman to break the sound barrier, flying at 840mph. (The first woman to break it? The aforementioned Jacqueline Cochran.) There was Bernice Steadman, who ran her own aviation school and would later subtitle her autobiography 'The Right Stuff – But the Wrong Sex'. There was Janey Briggs Hart, who at forty-one and the mother of nine was the oldest of the group, and soon to become the first woman in Michigan with a helicopter licence. And there was Wally Funk, who in 2021 would take a seat on a private sub-orbital flight funded by Jeff Bezos's Blue Origin company, thereby becoming, at eighty-two, the oldest woman to fly in space, but who now, at twenty-three, was the group's youngest member and who, on top of being a prize-winning competitive pilot, was also an expert marksman.

Cobb referred to them all as FLATS (First Lady Astronaut

Trainees); they also became known during the 1990s, courtesy of a television documentary, as 'the Mercury 13', even though those specific thirteen women passed in and out of the Lovelace Clinic separately and were never gathered together in the same place at the same time. Yet it's easy to conceive of them as a kind of parallel corps to the Mercury Seven, equally equipped for the mission, it seemed, yet excluded on a technicality from even applying to join.

However, despite the formidable claims of these women for serious attention, Lovelace could not get NASA to grant his 'Woman in Space' programme the official sanction which would have enabled him to take it further. Cochran's funding dried up, along with, it seems, her enthusiasm for the project, and the experiment halted.

Still, Jerrie Cobb had some momentum now. As a consequence of her extensive lobbying, a two-day Congressional hearing was tabled in Washington on gender discrimination at NASA, and she was summoned to give evidence before a sub-committee of the House of Representative's Committee on Science and Astronautics.* During testimony it was pointed out that when it came to flying in space, women actually had potential advantages over men. They tended to be smaller and require less room to move, which suited the cramped conditions of spacecraft, and they consumed less oxygen, food and fuel. In simulations, women had outperformed men in operating the controls for space rendezvous, and also in isolation and sensory deprivation tests, and they were less prone to heart attack. It was claimed, too, that women were less sensitive than men to radiation – a point which modern science would not support – and also to heat, pain, cold, monotony and

* This was two years before discrimination on the basis of gender in the advertising and allotment of jobs became illegal under law.

loneliness. (Modern science might not support all of that either, though judging by the phenomenon of 'man flu', it might support some of it.)

The women weren't trying to prove that they would be *better* than men at the job, even though some people seemed to need to set the bar for them that high. They merely wanted to prove that they *could* do it, equally well, and that they should be allowed to. As Cobb told Congress, 'We seek, only, a place in our nation's space future without discrimination.'

In Washington, Cobb and Briggs Hart tried to impress the sub-committee with reports of women's success under testing. To the startled disappointment of both of them, Jacqueline Cochran, former funder of the Lovelace programme, now stood up and spoke *against* their cause, declaring that 'men came first' to the space programme and noting that there was 'no shortage of well-trained and long-experienced male pilots to serve as astronauts'.

This was a switch which still causes puzzlement even now. It is speculated that Cochran had political and personal reasons for pulling the rug from under Cobb and the others, and a plan to institute a much larger women's programme at NASA later, under her own leadership.

Meanwhile, present on behalf of NASA were James Webb, the organisation's Administrator, and John Glenn and Scott Carpenter, shining stars of the Mercury programme whom the committee members seem to have been somewhat in awe of as the proceedings unfolded.

'I am not anti any particular group,' Glenn assured the room, 'I am just pro space. Anything I say is towards the purpose of getting the best qualified people, of whatever sex, colour, creed, or anything else they might happen to be.'

But then Glenn added the following assessment of the situation: 'Men go off and fight the wars and fly the airplanes and come back and help design and build and test them. The fact that women are not in this field is a fact of our social order.'

This remark became notorious, although in fairness to Glenn he did add a concession which is not often quoted: 'It may be undesirable,' he admitted. But there it was. Desirable or not, for him it was a fact, and in this matter the social order, in all its rigidity, had spoken, to the exclusion of all other arguments.

Webb, on behalf of NASA, made the point that this was 1962, that a Space Race was underway, and that this was no time to be messing around with experiments – which was a bit rich, given the fundamentally experimental nature of the entire space effort.

The hearing broke up without any recommendation for change, though the question of whether the military test pilot stipulation was sound did remain in the air for a while afterwards. Even then, though, it was clear that, for some people in positions of authority, just having to have this discussion was proving wearisome. The Executive Assistant to Vice-President Lyndon Johnson, Liz Carpenter, drafted a letter to NASA canvassing some views on the absolute necessity, or otherwise, of the military test pilot qualification, and put it in Johnson's in-tray for review. Johnson wrote across the top of it 'Let's stop this now'. The letter went unsent.

Years would pass before significant change occurred. Although NASA found it could easily suspend the military test pilot requirement for the fourth astronaut draft just three years later in June 1965, which was designed to bring some scientists into the fold (and yielded, incidentally, the arrival of six more men), that stipulation was immediately reinstated for future drafts, and would remain in place, entirely excluding women from the astronaut corps until 1978 and the Thirty-Five New Guys.

As so often, things moved differently in the Soviet Union.

In June 1963, the Soviet space agency was getting ready to launch its fifth orbital mission, Vostok 5, and the US could only wince and wonder at what special stunt their rivals would pull off this time. Russia's previous launch, in August 1962, had sent Vostok 3, piloted by Andriyan Nikolayev, into space for almost four days. The US's best duration at that point was just under five hours, so that was humiliating enough in itself.*

But that wasn't the end of it: while Nikolayev was still up there, casually clocking up his sixty-four orbits of the Earth, the Soviets launched Vostok 4, flown by Pavel Popovich, the folk-singing Ukrainian now getting the nod for a mission ahead of some of his Russian-born peers. Once in orbit, the two spacecraft flew to within 4 miles of each other and established radio contact – the first ship-to-ship communication in space.

A tandem launch! American hearts sank. What next? Loop-the-loops?

But the serious implication of all this was: if the Soviets could now manage two craft in space simultaneously, surely it would only be a matter of time before they were performing docking manoeuvres. And if they could dock spacecraft, it would not be all that much longer, clearly, until a Soviet crew was heading for the Moon.

Which is why, when news started to come through of another Soviet launch, all NASA personnel braced themselves.

At nearly 3.00 in the afternoon on 14 June 1963, Vostok 5, carrying Valery Bykovsky, blasted off and began to orbit the Earth. Two days later, Vostok 6 took off and joined it. So far, so familiar. Yet Vostok 6 was piloted not by another member of the vaunted Vanguard Six, nor by any other man in fact, but by someone called Valentina Tereshkova.

The two Vostoks orbited together and returned to Earth on 19

---

* Even by June 1963, when Vostok 5 launched, the best the US had managed was Gordon Cooper's one day and ten hours in Mercury 7 the previous month.

June, landing 200km apart and bringing with them, among other things, photographic evidence of the presence of aerosols in the Earth's atmosphere and scientific confirmation of female durability in weightlessness. There was no docking activity – NASA could be relieved about that. But the first nation to put a man in space had now trumped America again by getting a woman there.

A young woman, too. Tereshkova was just twenty-six. Her father had been a tank commander in the Russian Army and was killed fighting in Finland during the Second World War, when Tereshkova was only two. Her mother worked in a cotton mill, and Tereshkova, like her two sisters, left school at seventeen and went out to work, first in a tyre factory and then in a textile mill.

But she continued her education via a correspondence course and got a degree from the Light Industry Technical School. She also found time, between work-shifts and schooling herself, to become a competitive parachutist with her local sky-diving club. She seemed to like parachuting more than anything, and wrote in her diary, after her first jump, that she wished she could do it every day. The All-Union Voluntary Society had little hesitation in recommending her to the Soviet space agency.

Summoned to Star City and sworn to secrecy, Tereshkova told her mother she was off to train with the Russian parachute team. Quickly identifying her talent, Kamanin started referring to her with admiration as 'Gagarin in a skirt' and soon made her the subject of what from the outside, to those who had not witnessed her in training, must have appeared the boldest selection call in spaceflight history so far.

Tereshkova would always claim what she experienced in anticipation of her flight was not fear, but rather the kind of nerves that athletes know before a competition. At Baikonur, on the way to the launch pad, the bus stopped and Tereshkova got out to become the first woman to urinate against the rear tyre of a launch-bound crew vehicle, Gagarin's important tradition brooking no interruption by gender.

She had chosen the call-sign 'Chaika' – 'Seagull'. As the violence of the rocket's g forces gripped her, Tereshkova could be heard over the radio repeating that call-sign to herself like a mantra: 'Ya Chaika, ya Chaika' – 'I am Seagull, I am Seagull'. Inserted with a jolt into orbit, and abruptly afloat in an eerie silence, Tereshkova radioed down to Ground Control. 'It's me, Seagull. Everything is fine. I see the horizon: it is sky blue with a dark strip. How beautiful the Earth is!'

The two Vostoks passed within 5km of each other, and their pilots communicated over the radio, but stopped short of waving to each other – indeed, neither pilot could be sure they had made visual contact during their dual orbit. Still, over two days, twenty-two hours and fifty minutes, Tereshkova orbited the Earth forty-eight times, took photographs and monitored her spacecraft, and by the time she tore back into the atmosphere in a ball of flame she had been in space longer than all American astronauts to that point combined.

She ejected from the plummeting capsule, parachuted down in strong winds and made it to the ground, narrowly avoiding a lake and earning a bruised face somewhere along the way. The Seagull had landed. She then sat down for milk, bread and honey with some local villagers who came to her assistance and helped her out of her spacesuit.

Thirty years later, Tereshkova would report that her returning capsule had been locked into ascent mode rather than descent mode, and rapid adjustments had to be made from the ground to ensure that she didn't orbit away from the Earth irrecoverably and suspend herself in space fatally. At the time she was instructed not to mention this aspect of her adventure, and the mission was acclaimed an unalloyed success.

In her private post-flight debriefing, Tereshkova was able to highlight problems with the communications headset (Bykovsky also reported these), and to state on the record that she had vomited while trying to eat the supplied food. But she put this down not to any kind of problem with ingestion in weightlessness but to the taste of the food. She commended the ground crew on their thoughtful provision of toothpaste, along with her water, but she did upbraid them on their failure to pack a toothbrush.

Tereshkova also had an opinion to offer on American leaders and their attitudes to Jerrie Cobb and the Mercury 13: 'They shout at every turn about their democracy and at the same time they announce they will not let a women into space. This is open inequality.'

In Moscow, Bykovsky and Tereshkova received the full-bore heroes' welcome. With the former in military uniform and the latter in plain, dark civilian clothes, and with Gagarin beside them as usher, they were met by President Khrushchev and a display of a million flowers – at least according to the person doing the counting for the government-sanctioned broadsheet *Pravda*. What no one could dispute was that on that day, with the world's news cameras focused on her, Valentina Tereshkova, waving and smiling, was transformed directly from obscurity into nothing less than a global icon of female empowerment.

It was a role she enthusiastically embraced. All the cosmonauts were despatched on tours after their flights, but Tereshkova travelled more and further than any of them, frequently visiting and addressing women in factories and other places of work. As well as fulfilling a hectic schedule of appearances within the Soviet Union, she set off on forty-two trips outside it, including, in February 1964, to Britain where she accepted an award at the Piccadilly Hotel from the British Interplanetary Society and took tea with the Queen.

Five months after her flight, Tereshkova married her fellow cosmonaut Andriyan Nikolayev in a highly publicised ceremony at a government mansion in Moscow, hosted by none other than Khrushchev and attended by a rich mix of cosmonaut heroes and political heavyweights. Not exactly buying into the fairy-tale aspect of this betrothal, Kamanin quietly wrote in his diary that the marriage was 'probably useful for politics and science'. For the Russian population, though, the union of these two major space-stars, and now-confirmed Heroes of the Union, was a Soviet dream, making them the Beyoncé and Jay-Z of their day, but with an added layer of astrophysics and more medals.

Sadly, the relationship deteriorated, and in a way which was painfully public at times. Tereshkova grew visibly reluctant to pose with or even stand near her husband for photographs. The pair divorced in 1982. But the marriage yielded a daughter, Elena, the first person who could say that both their mother and their father had been to space. Elena graduated from medical school and became an orthopaedic surgeon.

As for Tereshkova, at the time of writing she is eighty-six and sits in the State Duma, the lower house of Russia's Federal Assembly. Tereshkova's standing as a legend of spaceflight and as a trailblazer for women in an overwhelmingly male-dominated realm is unassailable. She remains the youngest woman ever to have flown in space, and also, incidentally, the only woman to have flown in space alone, a record that is now unlikely to be taken from her.

Yet it would be nineteen years before the next Russian woman made it into space – Svetlana Savitskaya, who flew in 1982. And it would be another year after that before Sally Ride became the first female American astronaut.

It also needs to be noted that, although Russia galloped ahead in putting a woman in orbit, of the seventy-five women who at the time of writing have been in space, only five have been Russian. None of the other four in that original Soviet draft is among them. That daring piece of selection was not repeated in the Vostok and Voskhod eras.

It seems that in the pioneering days of spaceflight you could smash through the glass ceiling, only to have it instantly repaired in your wake, leaving the women who came behind you no choice but to take up the hammers again.

CHAPTER FOUR

GETTING OFF THE GROUND

'Hey, sky – take off your hat. I'm on my way!'

– Valentina Tereshkova, on the launch pad, 16 June 1963

I. ASLEEP AT THE WHEEL

Obviously I would have preferred it if, in the year leading up to my launch to the International Space Station in 2015, two separate rockets hadn't blown up shortly after leaving the pad and a third destroyed itself shortly after reaching orbit. It's tense enough sitting 50 metres above 150 tonnes of explosive fuel just before it ignites. You don't really need any recent dramas adding extra edge.

Then again, it wasn't like I didn't already know that getting blasted into the air on top of a rocket has its risks. Nobody ever tries to persuade you that it doesn't.

The good news about those three failures during 2014 and 2015 is that they were all attempts to launch cargo and, thankfully, involved uncrewed craft. But the third of those loads was a Progress vehicle, bound, like me, for the ISS. The rocket reached orbit and then failed to separate from the third-stage booster. The still-mated units went into an uncontrollable orbital spin lasting several days before dropping back into the Earth's atmosphere, burning up and disintegrating on the way through, and ending up at the bottom of the Pacific. If that malfunction had occurred with a crewed capsule attached rather than a load of cargo, the consequences for the astronauts on board were terrible to contemplate.

And that was the same configuration of Soyuz rocket, a Soyuz TMA-M, that sat under me on the launch pad at Baikonur a few months later.

Some calming historical perspective is always useful though. As rockets go, the Soyuz has a very strong pedigree. Dating from the sixties, it is the world's most used launcher, the veteran of more than 1,700 flights which have successfully put 140 crewed craft into space. To sit above a Soyuz rocket in 2015 took, without question, a different order of nerve from sitting above a Saturn V in 1968, as the Apollo 8 crew did. At that point, Frank Borman, James Lovell and William Anders were about to be blasted further than anyone had ever travelled (around the Moon, of all things) by the heaviest and most powerful rocket ever devised – and which had only been fully tested twice before, and never with humans on board.

Which must have called for some trust, to say the least.

By contrast, by the time I got to the top of it, the Soyuz's robust technology and its ability to launch in all weathers had already enabled the Russian and Soviet space agencies to taxi a steady flow of crews to the Salyut and Mir space stations for three decades, and to continue to ferry astronauts and cosmonauts to the ISS. If you were selling a Soyuz second-hand on WeBuyAnyRocket.com, you would be stressing its reliability: good runner, always starts on cold mornings, one careful owner, etc.

One can confidently state that Soyuz rockets work, then.

Except on the occasions when they don't.

In September 1983, Vladimir Titov and Gennady Strekalov sat on the launch pad at Baikonur aboard, in their case, Soyuz 7K-ST No.16L, to give the flight its full designation, or Soyuz T-10a as it also became known. They were bound for Salyut 7, the last of the series of six space stations which the Soviet Union put into low Earth orbit between 1971 and 1986. (The Salyut series begat the Mir space station, which in turn begat the multi-partner ISS.)

The launch seemed to be proceeding normally until ninety seconds before ignition, when a valve failed in the propellant line and flames suddenly erupted around the rocket's base.

At that point, it would have been clear that the rocket was extremely likely to blow up, 150 tonnes of kerosene detonating in a massive conflagration. And there would probably only be a handful of seconds before it did so.

Fortunately for Titov and Strekalov, this was exactly the kind of emergency for which the automatic Launch Escape System was invented. The Soviet space programme had come on a long way by now from the rickety escape plans involving nets, ropes and bathtubs that had been conceived around the earliest launches. The LES was primed to fire automatically in the case of an abortive launch, prising off the capsule and shooting it high into the air, like a cork coming out of a champagne bottle, and, hopefully, flinging the crew on board to safety. There were parachutes for the capsule's descent – and not a bathtub in sight.

Alas, in this case, the flames engulfing the rocket burned through the command cables before the system could be triggered – fire maliciously destroying the very arrangement meant to spare the cosmonauts from the effects of fire. As the flames grew, the capsule stayed exactly where it was. A full twenty seconds after the fire was detected, the LES was finally activated by radio command by controllers in the panicked block-house.

The capsule broke free from the rest of the rocket and shot skywards. Seconds after that, the stricken rocket's fuel tanks exploded. The fire on the pad was to burn for twenty hours.

In the violence of their ejection, Titov and Strekalov were

thrust upwards with such force that, for a period of about five seconds, they were subject to the almost concussive pressure of 17 g. The capsule soared to 2,000 feet, then began to plummet, until its parachute popped, slowing its descent back to the ground.

Even at that point, though, the crew's fate was uncertain, dependent on where the capsule landed, and how hard. Years later, in an interview for a documentary on the History Channel, Titov jocularly maintained that he'd had the presence of mind to deactivate the capsule's cockpit voice recorder so that he and Strekalov would be free to swear as freely as they felt they needed to in the final stages of their extremely rough journey.

One wonders, though, whether he had a bad feeling about how this was going to end, and whether his motive in stopping the tape was to avoid leaving a dire recorded legacy akin to the one unavoidably left by the Apollo 1 crew. Maybe Titov used the off switch to spare his and Strekalov's loved ones.

Anyway, in the event, the pair hit the ground hard but safely on the scrubby, flat Kazakh steppe, 4km from the pad. Titov and Strekalov had been in the air for five minutes and twenty seconds, and were winded and badly bruised but otherwise in one piece.

The rescue helicopter reached them almost immediately and its crew went to the hatch.

'Are you OK?' one of the team asked.

'Do you have a cigarette?' Titov replied.

The crew did. Titov and Strekalov lit up.

I can't say that I have ever been a smoker, but I am reliably informed that the best cigarettes are (1) after dinner, (2) after sex and (3) after surviving a 17 g launch abort from an exploding Soyuz rocket on the pad at Baikonur.

Titov and Strekalov must have wondered whether they were cursed. Their previous mission together, Soyuz T-8, just five months earlier in April 1983, was intended to dock with the

Salyut 7 space station. When their craft got within range of the target, however, its radar antenna, intended to guide it during the automatic docking, wouldn't deploy. (It seems to have been struck and damaged when the rocket's fairing was shed during launch.) Titov was required to attempt a manual docking using a not particularly revealing optical sight on the control panel and radar inputs from the ground.

Just to complicate things further, they were in almost complete darkness.

The craft was on its final approach when Titov became convinced that he was coming in too fast, braked to avoid a collision with the wall of the space station, and was then obliged to back off and pull out altogether. It was deemed too risky to make another attempt. He and Strekalov turned the capsule around and returned disconsolately to Earth.*

So that was two consecutive missions for Titov and Strekalov that had ended prematurely. Nevertheless, the old wisdom about getting straight back on the horse after a fall clearly applies to rockets, too. Strekalov was to fly to space four more times after that bruising launch escape, and had the distinction of serving missions on three different space stations.

As for Titov, he went on to serve a long-duration mission on Mir, and to fly as a Mission Specialist on Space Shuttle *Discovery*

---

* As I recorded in my book *Limitless*, at the end of my journey to the ISS, Yuri Malenchenko was required to perform a manual docking when a thruster failure caused the abort of our automatic approach. It took him two goes, peering, like Titov, into a mostly dark periscope and wrestling all the while – again, like Titov – with the possibility of colliding with the ISS and being sent spinning off into space. But he managed it. I could only sit there and spectate. Those were without question the longest minutes of my life. They were also minutes in which I contemplated, among other dire outcomes, the possibility that we, like Titov and Strekalov, would have to turn round and go home, my mission over before it had begun.

in 1995, on STS-63, in which *Discovery* rendezvoused with Mir – a coming together of nations that would have been unthinkable in the fifties and sixties. Titov also flew to Mir on Space Shuttle *Atlantis*, in 1997, and by the time he retired, in 1998, he had spent a total of nearly nineteen hours walking in space.

Meanwhile, the pair remain the only crew ever to have used the Launch Escape System with the rocket still on the pad. I suppose we can chalk that up as another historic Soviet spaceflight first. By rights, though – and as so often in the Space Race story – Titov and Strekalov should have been beaten to that place in the record book some time before by the Americans.

Eighteen years before, to be precise. In December 1965, Wally Schirra, from the Mercury Seven, and Tom Stafford, from NASA's second draft, were sitting on a Titan II rocket at Cape Canaveral, awaiting the launch of Gemini 6A, commissioned to fly up and rendezvous with Gemini 7, already in space for a long-duration mission lasting a fortnight and crewed by Jim Lovell and Frank Borman. This was to be a critical test of the US programme's ability to bring orbiting craft together and, accordingly, a major step on the way to a Moon mission.

When the countdown reached zero, the rocket's engines ignited. One and a half seconds later, however – and just 1.7 seconds before expected lift-off – for no immediately discernible reason, the engines completely shut down.

In the capsule, the mission clock was now running, and Schirra, as Command Pilot, knew what this meant: he was aboard a launched but abortive rocket, and he should immediately reach down and pull the orange-coloured D-ring situated between his legs, thereby detonating the ejector seats and flinging Stafford and himself out of the capsule and away. If a fully fuelled rocket lifts even slightly off the pad and then thinks better of it and drops back down again, the risks of a catastrophic explosion are high.

But Schirra didn't do that. He stayed his hand. Somehow, in that split second, and amid the noise and ratcheted tension of launch, the astronaut made the judgement that he hadn't felt the spacecraft lift at all before it shut down and that therefore he and Stafford were still safely on top of a non-firing and fully moored rocket that wasn't about to explode and kill them.

Not only did Schirra thereby spare himself and Stafford a bruising and dangerous eviction from the capsule, he also saved the mission. Ejecting would have cost days of delay while the spacecraft was made flight-ready again, and would have ruled out the possibility of the rendezvous with Gemini 7. As it was, Schirra and Stafford were able to launch three days later and attempt to dock.

Sometimes the best thing to do is nothing. Schirra's call that day remains perhaps unmatched in the history of spaceflight as a demonstration of cool on the launch pad. By his own later estimate, it also saved his and Stafford's lives. In 1997, Schirra would relate what he had belatedly realised: that by the time of the aborted launch, he and Stafford had spent one and a half hours in the pressurised Gemini capsule, inside their spacesuits, essentially soaking themselves in pure oxygen. The spark from the explosive ejector seats would most likely have ignited the astronauts (as Schirra put it) 'like Roman candles'. Nobody seemed to have considered this possibility because the Gemini evacuation system had only been tested in a mock-up capsule with a nitrogen atmosphere, in inert conditions. Not until the Apollo 1 fire just over a year later would NASA fully wake up to the perils of putting flammable materials inside oxygen-rich spacecraft.

Schirra and Stafford narrowly dodged death on that occasion, but it's Vasily Lazarev and Oleg Makarov who hold the unique double-distinction of surviving the physical stresses of an automatic launch abort *and* flirtation with a cliff-edge.

The two cosmonauts had taken off from Baikonur on Soyuz 18a on a clear morning in April 1975, on a mission to dock with the Salyut 4 space station. Four and a half minutes after ignition, Lazarev and Makarov were 145km up in the air waiting for the six explosive bolt-locks to blow and cause the rocket's second and third stages to separate and fall away. But only three of those locks did so. The remainder held, causing the second stage to cling on until the third stage ignited and broke free. The rocket began to lose power and tip over.

At that point, the launch abort system activated and blasted the capsule free. However, at the moment it did so, the rocket was pointing downwards. Lazarev and Makarov were shot out on a ride whose accelerative force, according to some estimates, reached as high as 21.3 g. Fortunately, the violence of the descent did not prevent the parachute system from working. The capsule eventually struck the ground in a survivable impact.

Their adventure wasn't over, though. Lazarev and Makarov had landed, inconveniently, on a remote mountain range, on a steep snowy slope. Even more inconveniently, the steep snowy slope led to a sheer drop onto rocks below.

With grim inevitability, the capsule now began to slide downwards through the snow and ice, with Lazarev and Makarov bruised and oblivious inside. In a development that seemed to have come directly from the plot of a Roadrunner cartoon, it appeared they had survived a 145km fall from the sky only seconds later to plunge off the edge of a cliff.

But once again they had their parachutes to thank. As the capsule slithered towards the cliff-edge, dragging the parachutes behind it, the fabric and cables snagged on some rocks and halted its slide.

Lazarev and Makarov climbed out and found themselves ankledeep in snow and in temperatures that, over the next twenty-four hours, would slump as low as minus 7°C. But where exactly were they? It occurred to them that they were quite likely in China. Just in case, they cleared from the capsule all the papers relating to a military experiment which was part of their mission and burned them on the fire they had built.

In fact they were in the Altai mountains, 500 miles inside the Russian border. But better safe than sorry.

They were there for a day and a night before the rescue team reached them. Soviet reports of the incident were typically sketchy but on this occasion, because relations had considerably thawed and planning was underway for a collaborative Apollo-Soyuz project, the US felt able to request more information. In the report they received, the episode was referred to as 'the April 5th Anomaly' – a splendid piece of positive thinking which became the incident's label for years afterwards.

Makarov flew three further Soyuz missions. Lazarev, however, who was initially reported unhurt, had sustained injuries in that capsule expulsion and the subsequent landing which were serious enough to prevent him flying again. He died in 1990 at the age of sixty-two.

At the time of my launch, Soyuz 18a was the only crewed Soyuz flight to have undergone a launch abort at high altitude. But it happened again, three years after I flew. In October 2018, Nick Hague, from NASA's 2013 draft, and the cosmonaut Aleksey Ovchinin, with whom I spent time on the ISS, launched from Baikonur on Soyuz MS-10. The rocket left the pad and climbed into the air. And then, with their families and friends watching in horror from the ground, it seemed to become a splattering of distant white fireworks against the clear blue sky.

During separation, the launcher's first and second stages had collided. One of the fireworks that the spectators were seeing was actually Hague and Ovchinin being evacuated, hurtling away from the rocket and leaving a white trail behind them. The capsule's automatic abort shot them on a 7 g ride all the way to 93km, just below the Karman Line. A little over a quarter of an hour later they floated back down to the ground, landing on a blessedly cliff-free patch of Kazakhstan 250 miles from the launch site. Both of them were unharmed and flew back to Baikonur to be reunited with their traumatised families.

So, yes, as I was saying: the Soyuz is a completely reliable mode of transport, except on those occasions when it fails to be. And, clearly, it would be wisest not to spend the evening before your flight flicking through the Bumper Book of Soyuz Launch Failures – which, it's again worth stressing, would not be a particularly bumper book in any case.

But would any of these tales of malfunction have dissuaded me, or any other trained astronaut, from climbing into one? Clearly not.

The launch pad is where the astronaut dream finally stops being anticipation and starts to become real. After years of rehearsal, it's the stage on which your career's main act begins. Of course, as stages go it's an unusually volatile and dangerous one. People will ask, naturally, 'Were you afraid?' But most astronauts, I think, would tell you that they had done their bartering with fear well in advance of launch day, and that what is mostly in your system at the point when the bus draws up alongside the rocket is raw adrenaline. Certainly there are no recorded instances of any astronaut having second thoughts on the pad and asking to come down and go home. It goes without saying that by this point you're all in, or you wouldn't have reached this point at all.

When I think back, anxiety was certainly a big part of the run-up to my launch. But it wasn't anxiety about the launch itself. I was anxious about some unforeseen delay or unavoidable truncation happening to the mission, anxious about something showing up in the seemingly endless last-minute medicals, anxious about waking up with the flu, anxious about twisting an ankle while skating with my sons during our last spell of proper time together in Star City before my statutory fortnight in quarantine. (You've never seen anyone shuffle so stiffly and carefully round a frozen pond as I did on that occasion. I was practically on all fours.)

Altogether, I was so preoccupied with dread thoughts of something randomly intervening at the very last minute and curtailing or even snatching away entirely my long-awaited and deeply longed-for journey to space that nothing else got a look in.

Still, I can only wonder at the pioneers of the Mercury and Vostok programmes who flew solo and lay in those capsules on the launch pad on their own. On top of that Soyuz rocket, I was shoulder-to-shoulder with two flown astronauts: Tim Kopra, already the veteran of an ISS expedition from which he had returned to Earth aboard the Space Shuttle *Discovery*, and the highly experienced cosmonaut Yuri Malenchenko, the definition of a calming presence. Yuri was making his sixth trip to space and he exuded so much assurance, not just inside the spacecraft but everywhere he went, it was hard not to catch some of it just by being near him.*

There was no reassuring companionship at the vital moment for the Mercury and Vostok pilots – for the pioneers, the likes of John Glenn, Wally Schirra, Gherman Titov and Valentina Tereshkova. In those final minutes ahead of a ride which – assuming all went well – would take them from a standing start to the

---

* Yuri was so comfortable in space he got married there. He was betrothed to Ekaterina Dmitriev in a ceremony on the ISS in August 2003. Ekaterina was in Texas at the time; Yuri was somewhere over New Zealand. Their wedding day was already in the diary when Malenchenko's mission got extended, but they went through with it anyway, Yuri appearing in Texas in the form of a life-sized cardboard cut-out. Theirs were the first space-based nuptials, but as commercial sub-orbital spaceflight takes off you can expect there to be many more in the future.

practically unimaginable speed of 17,500mph in around nine violent, loud and unquestionably precarious minutes, they had only themselves for company.

Each of them will have watched the last member of the ground team withdraw, felt the hatch close, heard the bolts go in and then found themselves entirely alone – the only exposed human at the centre point of a launch zone rigorously cleared and made safe to a radius of 3 miles all around them. It's hard to believe that every nerve in their bodies wasn't thrumming at that point, or that they had ever felt more alert.

Which only makes even more remarkable what the ground crew heard as a fully fuelled Atlas rocket sat on the pad in May 1963 awaiting the launch of Mercury 9, the last of the Mercury solo orbital flights. Over the intercom from the capsule came the sound of Gordon 'Gordo' Cooper . . . softly snoring.

Cooper's launch had already been scrubbed once, the day before. That time, when he was assisted into the capsule, he found that Alan Shepard had left a departing gift for him on his seat: a small suction-pump labelled 'Remove Before Flight', wryly honouring the fact that Cooper would be the first astronaut to fly with a urine-collection device inside his spacesuit.* Hours of delay followed before the flight was postponed and Cooper was unstrapped and extracted from the capsule, remarking casually to the ground staff on his way out, 'I was just getting to the fun part.' He got changed, climbed in his car and drove off to spend the rest of the day fishing.

And the day before *that* he had almost lost his seat in the capsule altogether by deciding, just for fun, to repeatedly buzz the

---

* The Mercury Seven seem to have enjoyed leaving each other pre-flight gifts. John Glenn hung a sign over the controls in the capsule that carried Al Shepard to space. It read: 'No Ball Games'.

Cape Canaveral Administration building in an F-106 jet – the building, I should add, where, at that very minute, the Project Mercury Operations Manager Walt Williams was trying to hold a pre-launch meeting with senior NASA staff. Apparently there were people at that meeting who thought this behaviour was not only unamusing but also grounds for Cooper's immediate removal from the mission. Watching his colleague noisily dive-bombing a significant segment of the space agency's upper management, Al Shepard, the appointed reserve astronaut for Mercury 9, was confident that he was about to get promoted to another flight.

Yet, by the skin of his teeth, Cooper survived – and, indeed, once nudged from his slumbers on the launch pad he flew a great mission, bringing the capsule back to Earth under manual control in the face of a cascade of electrical problems, at the peak of which he produced the supremely calm and now legendary line 'Things are starting to stack up a little here'.

He was in space for just over thirty-four hours and thereby had the honour of being the first American to sleep not just on the launch pad but also in orbit. During his scheduled sleep, medics observed a significant but only momentary increase in Cooper's heart-rate – evidence, it was concluded, of an exciting dream. The astronaut never related that dream's contents, or whether he recalled them. But only in sleep, it seemed, did Cooper become at all ruffled. He was the last astronaut to fly a completely solo orbital space mission, the last of the true loners.

Cooper was by no means the last astronaut to fall asleep on the pad, though. While Apollo 17 idled on the launch pad, forced into a two-hour delay, Ron Evans, the Command Module Pilot, dozed off. Three hundred feet up a rocket, wearing a stiff pressurised suit and lying on his back next to two colleagues, he had doubtless known more comfortable beds. And yet . . . Beside him, Gene Cernan and Harrison Schmitt grumbled to Mission Control about the snoring. No spare room to slide off to in this particular situation.

And then there was Robert Crippen, the pilot of STS-1, the maiden flight of the Space Shuttle *Columbia*, in 1981, the most ambitious launch of a spacecraft since the Apollo era and a massive challenge for its crew. The launch was held up and, by his own admission, Crippen 'dozed off and went to sleep a few times'. During one of those dozes his slumbering arm dropped and bumped against a control panel, knocking the plastic top off a switch. The launch delay provided time to perform a quick fix.

Marsha Ivins flew on the Space Shuttle five times between 1990 and 2001, so she had more chances than most to get used to the routine. 'The truth is,' she once pointed out, 'there isn't much to do for those two hours after you climb in. Many astronauts just take a nap. You're strapped in like a sack of potatoes while the system goes through thousands of pre-launch checks. Occasionally you have to wake up and say "Roger!" or "Loud and clear!".'

Leroy Chiao, who flew three Shuttle flights and was the Commander of Expedition 10 on the ISS in 2004–05 (when, incidentally, he became the first American citizen to vote in a presidential election from space), thought of the pre-launch phase as 'time to relax a bit. The environment is totally familiar, thanks to the hours upon hours spent in the simulators. For once nobody is talking to you. Nobody is asking you for something. It's not unusual to doze off.'

Gunter Wendt, the American-born German engineer who worked on the Apollo programme, stalking the Space Center in his bow-tie, steel-frame glasses, white cap and white lab coat, once gave an interview in which he compared the phenomenon of the sleeping astronaut with that of the conked-out holiday-goer: you stress in advance about the alarm clock going off, stress in advance about getting to the airport on time, stress in advance about the queues at security ... And then, once you're on the plane and sitting back in your seat, the abrupt release of all that stress causes you to drop off, even before take-off. 'And that's about the same as these guys,' Wendt suggested. 'They have been through so many tests, so many dry runs, so many activities, that finally they say: "Oh man, just close the hatch and let it go."'

On top of that Soyuz rocket, I didn't fall asleep. When the ground team, just for their own amusement, piped Europe's 'The Final Countdown' into our capsule, I maybe wished I *had* been asleep. But I can see how sleep happens.*

That was in the pre-launch downtime, though – the lull before the storm. Asleep *during* launch? Impossible. Even for someone whose veins ran as chilled as Gordon Cooper's.

When that rocket went up, I had never been more awake in my life.

II. GOING UP

'Ten ... nine ... eight ...'

... is what you *won't* hear, in fact, as you sit there in the capsule. Or not if you're launching in a Soyuz from Baikonur anyway. No countdown in your ear – that's for Ground Control. For you, it's the voice of the instructor, calmly announcing the launch sequence's final stages, all the way down to 'Engines at full thrust'.

---

* Each of us was allowed to choose three songs to be played in that last hour. I picked U2's 'Beautiful Day', 'A Sky Full of Stars' by Coldplay and 'Don't Stop Me Now' by Queen. I also got them to throw some Lady Gaga into the mix because I knew it would irritate Tim Kopra. Yuri didn't choose anything. I guess when it's your sixth trip to space you run out of reasons to lean on U2 for support, but I was grateful for their backing.

But we're getting ahead of ourselves. This is how the day of your launch will most likely have panned out so far.

First, you will have had a visit from the flight surgeon – the final medical and weight check – prior to an anti-bacterial shower and enema followed, somewhat contradictorily, by a hearty breakfast (eggs, porridge, pancakes, gallons of black tea), followed by a blessing – including a thorough drenching with holy water – by a Russian Orthodox priest.

Then, wearing a flight suit over sterilised white long-johns, you will have boarded a bus and driven for thirty minutes to Building 254, where you will have been installed in your carefully pressure-checked Soyuz spacesuit, with your friends and invited guests watching through a large glass window. And then, with grey Wellington boots on and carrying with you, like a doll's suitcase, the box that hosts your suit's cooling unit, you will have reboarded the bus and driven to the launch pad.

This is the bus, incidentally, which will pause on that journey so that you can climb down, briefly unfasten that carefully sealed and sanitised Soyuz spacesuit, and empty your bladder in memory of Yuri Gagarin. Then you will seal yourself up again as best you can and climb back on the bus.

At the pad, you will take the elevator to the rocket's top floor, 50 metres up, observing, as you go, the frosting on the rocket's side and the clouds of vapour already pouring off it, and becoming intimately aware of its scale and its pent-up power.

With one last, deep intake of fresh air, and one last look across the Kazakh steppe and down to the crowd of onlookers way off in the distance that contains your family and friends, you will enter the rocket, assisted by the ground crew, and squeeze down through the habitation module, loaded tight with cargo, into the descent module, and find your seat, taking extreme care, of course, not to knock against any switches.

You will plug in two electrical cables, one connecting your headset, the other connecting the medical harness beside your chest which will monitor your heart-rate and breathing throughout the flight. You will plug into your suit the ventilation tube, which is your personal aircon, and the oxygen tube, which is for emergencies, and you will lean forward and attach the knee-braces. And then finally, very tightly – as tight as it will go – you will strap yourself into the five-point harness.

There's a long list of pre-flight checks to be run through now, and you and your crewmates will likely be busy for the next fifty minutes or so. And there's actually something quite peaceful and even soothing about this phase, after the hectic and frequently noisy fortnight leading up to the launch – the ceremonies, the press duties, the farewells, formal and informal. The hatch is closed and now it's just you and your team, quietly getting on with the job that you've been trained to do.

But then it's about sitting back and waiting for the moment.

The noise comes from way below you at first, but seems to advance on the capsule until it is right by your ears. It starts out as a rumble and builds quickly to a deafening roar. You feel the rocket sway slightly from side to side, as if buffeted by gusts of wind, and then you feel it lifting, almost imperceptibly at first but then unignorably, slowly and steadily. And the noise rises with you, the roar still growing and broadening but now with an added crackle, vast in volume.

With height comes speed, the rocket accelerating rapidly up to 4 g, pinning you to the back of your seat while you clench your stomach muscles against the worst of its effects, as you've been trained to do. There's a bang, which is the sound of the Launch Escape System getting jettisoned, and then, on two minutes, a major jolt as the four first-stage boosters separate.

With their loss you feel a rapid deceleration and form the

faintly alarming impression that the rocket is falling, dropping back in the direction you just came from, which would be . . . not optimal.

But then the second stage kicks in and the acceleration begins again, more smoothly this time, pulling you higher and higher, up to a relatively easy-going 1.5 g.

You may well be oblivious to the moment in your flight when you officially become a flown astronaut. Obviously there are no signs to look out for: 'Entering Sub-orbital Space – Please Drive Carefully Through Our Karman Line'. Nor does the cockpit automatically light up and a fanfare sound, as on a jackpot-winning Las Vegas fruit machine. In rockets, you're generally in favour of cockpits *not* lighting up and fanfares *not* sounding.

Nevertheless, that exact point in the voyage seems to have been a big moment for Mike Mullane's crew aboard Space Shuttle *Discovery* in 1984, with a lot of whooping as 100km ticked by, and some dark Shuttle-related humour about how if the thing blew up now that would be fine because at least the rookies would die as flown astronauts, with gold-wing badges to their names rather than silver ones.

But clearly that was a Shuttle thing. Aboard the Soyuz you will be passing through the Karman Line merely seconds after the rocket's nose-fairing splits and falls away, finally unblocking the windows, meaning that you will be getting your first tantalising, imperfect glimpse of space's approaching blackness, and therefore potentially be too distracted to note the altitude.

And if the window doesn't distract you, maybe another huge jolt will. That's the second stage decoupling, leaving only the third stage propelling the spacecraft.

Which means it's time for the properly speedy part of the ride.

The rocket now accelerates back up to 4 g, but you are flying horizontally rather than vertically, and so now you feel as though you are being blown forward by an astonishingly powerful wind. The sensation of pure velocity runs right through your body. The thrust is steady, but the craft is burning through its fuel load and rapidly losing weight, so that its pace quickens and quickens, rushing you onward in an ever-gathering surge of power which will continue quickening for four minutes, carrying you along to 4.9 miles per second and forcing you to where gravity can no longer hold you.

And just as you are asking yourself, in astonishment, how this much acceleration can be possible for this much time, you will be hit by another massive jolt, the strongest, full-body blow of the journey, as the third-stage engine cuts out, flinging you forward against your straps as you go from 4 g of acceleration to zero as abruptly as if you had flown nose-first into a wall.*

And suddenly, everything is still and quiet – eerily so after all that amazing noise and agitation. And anything not strapped down in the capsule, such as a pen left on top of the control panel, and possibly one or two loose screws, will rise slowly and begin to float in front of your eyes. Eight minutes and forty-eight seconds have passed and you are 200km away from Earth and in orbit – and, yes, most definitely an astronaut.

At that point you are free to loosen your straps a little so that you too can float up slightly and get a view out of the window. And if, like me, you are lucky, at the exact moment you look out you will see the Moon rise over the Pacific, its silver disc emerging not gradually but all at once, like a light going on, while the blue Earth shimmers below it. And it will be one of the most amazing

---

* The equivalent moment in the flight of the Space Shuttle appears to have felt very different. Robert Crippen has recalled how, on the maiden *Columbia* flight with John Young, a veteran of Saturn V flights, both had preemptively braced themselves with their hands against the windscreen, waiting for the jolt at the cut-out. But it never came. 'I never felt it,' Crippen said. 'And I don't think John did either.' The Shuttle seems to have glided into orbit.

sights you have ever seen – something, indeed, it occurs to you, that you would risk everything to see.

Which, as it happens, you just did.

III. 'EVERYTHING IS NORMAL'

Two days before his launch, Yuri Gagarin sat in the cottage where he was lodging near the Baikonur launch site and wrote a letter to his wife, Valentina, and their young daughters, Elena and Galina. Then he sealed it and tucked it into his suitcase where he knew they would find it when his things were returned to them in the event of his death. Addressing the letter to his 'sweet and much loved Valechka, Lenochka and Galochka', Gagarin excitedly revealed that he had been selected to be the first man to fly to space. He wanted his wife to be happy for him. Happy that 'a simple man' had 'been trusted with such a big task for his nation – to blaze the trail!'

He explained that the launch would be in two days' time. He imagined them all at home at that time, going about their normal lives. He told them how much he wished he could be with them before the flight, even for just a short time, and talk to them.

He then told them how much he trusted the machinery that would take him to space, insisting that it would not fail. But he had to be realistic with them, too: accidents happen all the time, he wrote. And if an accident did befall him, then he wanted them all – the children, and Valechka above all – 'not to waste yourself with grief'.

He asked his wife to take good care of their daughters. 'Love them l like I do.' And to make them worthy of the Soviet state.

As for her personal life, he told her that she had no obligation to him, that he had no right to insist on anything.

And then he seems to have worried about the tone he was striking – that he had become 'too gloomy'.

He finished the letter by quoting some words he had read in childhood by Valery Chkalov: 'If being, then be first.'* That had been his motto, and would be until the end. In the meantime, he dedicated his upcoming flight to the Soviet people and to communism, 'to our great motherland, to science'. He signed off the letter: 'This seems to be all. Goodbye, my dears. I embrace you all tightly and kiss you. Your dad and Yura. 10 April 1961. Gagarin.'

He was a twenty-seven-year-old military pilot who had never travelled beyond the boundaries of the Soviet Union. And two days later he would attempt to travel beyond the boundaries of Earth.

Gagarin didn't sleep much the night before. That was partly the adrenaline and partly the fact that the medical team had wired him up before bedtime so that they could monitor him through the night. They had done the same with Gherman Titov, who was acting as reserve cosmonaut (and who was still, even at this late stage, hopeful of an opportunity to step in), with the result that both of them lay tensely awake, barely daring to move in case their read-outs were deemed by the medics to show them unfit for the mission in some way.

In the morning, the ground crew helped him into his orange flight suit then gathered around him, offering him blank pieces of paper.

Gagarin was puzzled. Paper?

And then he realised: they wanted him to sign them. Nobody

* Valery Chkalov died in 1938, crashing a prototype Polikarpov I-180 fighter plane during a test flight, when Gagarin was only four. But he was a legendary and inspirational figure in Soviet aviation, breaking records for long-distance flying and also once managing to pull 250 loop-the-loops in a bravado forty-five-minute session.

had ever asked Gagarin for his autograph before. It was something he would have to get used to.

Following this impromptu signing session and just prior to the team boarding the blue and grey bus for the launch pad, somebody realised that, although he had been carefully supplied with a hunting knife and a revolver, Gagarin had nothing about him to show that he was a Soviet cosmonaut. No patch on his suit, no sign on his helmet. In the very likely event of him parachuting to the ground in a remote spot, he might easily be mistaken for a downed American spy – the same fears, in other words, that had been in play when our friend Ivan Ivanovich, the slightly-too-lifelike mannequin, had set off for space just over a fortnight earlier.

So a pot of red paint and a brush were quickly drummed up, and the legend 'CCCP' was applied to Gagarin's white helmet by one of the engineers, Gherman Lebedev. Gagarin then travelled the final metres to the spacecraft with the still-wet logo on his head.

Even by the hurried standards of the Soviet space agency during the Space Race, when things still technically under construction quite often found themselves forced into active service, this omission seems remarkable. So much thought had gone into the specifically Soviet nature of this potentially historic moment and its resonances for the rest of the world that it's hard to imagine care not being taken to ensure at least some kind of highly visible Soviet branding on Gagarin's person.

Yet there are photographs of Gagarin, fully suited, pre-launch, with a blank white helmet – and with his typical huge smile. And then there are photographs of him on the bus to the launch pad, looking more pensive, and film footage of him heading across the tarmac in the last moments before entering the capsule, in both of which his helmet has acquired four very neatly painted if not quite regular letters. The helmet now resides in the Memorial Museum of Cosmonautics in Moscow and you can still see how the letters are not quite uniform and how the colour is irregularly applied.

No drips or smudges, though. Hats off to Lebedev for his steady hand.

The farewells at the foot of the rocket were protracted. Titov, who was also fully suited up, went to give Gagarin the traditional Russian triple-kiss and, in an absurd moment of space-age chivalry, the two of them ended up repeatedly clashing helmets. Other members of the Vanguard Six, who had travelled on the bus in their pilots' leather jackets, also embraced Gagarin and bumped their unprotected heads in the attempt. Andriyan Nikolayev claimed to have had a bruise on his forehead afterwards.

Nikolai Kamanin, fighting back his anxiety, shook Gagarin's hand and said, 'See you in a few hours.' He knew that the odds of that were, at best, fifty-fifty. Of the previous twenty-four tests of the Vostok launch system, twelve had failed, many catastrophically.

Then again, twelve had succeeded. You had to look on the positive side.

Eventually the sea of well-wishers parted and Gagarin ascended a flight of steps before turning to wave, his gloves dangling on ties from the wrists of his suit. And then he disappeared into the lift. It was around 7.10 in the morning on 12 April.

It was the duty of the engineer, Oleg Ivanovsky, to ensure Gagarin was safely strapped into the capsule and then to tighten the thirty bolts which would close and seal the hatch. As Ivanovsky prepared to leave, he whispered something into Gagarin's ear, moving in close so that the intercom wouldn't pick him up.

He told him that he had the secret code numbers for the manual override – the ones hidden somewhere in the cockpit in a sealed envelope – and that those numbers were 1–2–5.

Gagarin laughed and told Ivanovsky he knew them already. Kamanin had quietly slipped them to him earlier.

And so, actually, had Mark Lazarevich Gallai, a Soviet test pilot and key member of the cosmonaut training team.

And so, actually, had Sergei Korolev, the director of the Soviet space programme, who referred to the Vanguard Six as 'my little eagles' and made no secret of his particular fondness for Gagarin.

So much for the protocol around the ultra-confidential code in the hidden envelope.

Gagarin was in the capsule for almost two hours while the pre-flight checks proceeded. At certain points he could be heard quietly singing to himself. He did not sleep.

Finally it was time. From the control room, Korolev told Gagarin, 'Have a good flight,' and then added, 'Good luck.'

The four first-stage engines ignited, pouring streams of flame into the pit below them. The support gantries tilted back and, through billowing clouds of steam, the rocket tremblingly began to rise off the pad.

Inside the capsule, Gagarin let out an exhilarated whoop. 'Poyekhali!' he shouted. ('Let's go!')

As the rocket lifted away from the ground and began to gather speed, his heart-rate rose from a languid 64bpm to an urgent 157bpm.

'I heard a whistling and increasing noise,' Gagarin later recalled. 'I felt how the whole body of the giant spaceship started to tremble and slowly, very slowly, took off. It was very hard to move my arm, but I knew this condition would not last very long, only while the ship acquired the necessary speed.

'The Earth reported 70 seconds have gone since take-off. I replied, I understand – everything is going well.' He had chosen the call-sign 'Cedar'. On the ground, they heard Gagarin say, 'This is Cedar. I am feeling fine.'

Two minutes in, the opening salvo complete, the first-stage

engines were jettisoned and fell away and now the second stage began. Only now did the payload fairing fall away, granting Gagarin a view from his window. Not that he was able to take anything in yet, forced back into his seat as he was by the rebuilding g forces, his vision blurred by the craft's violent juddering as the second-stage engines forced it onwards and upwards.

Everything was going exactly to plan. Rising tension still thrummed through the control room though, and it would climb to its peak seven minutes into the flight, at the point at which the third-stage engine started to do its business, firing to give the craft the final push into orbit.

It was the third-stage engine that had given them all the trouble up to now. Everybody knew that if there was going to be a catastrophe on the way up, it would most likely come here.

And sure enough, Korolev, to his horror, could now see descending numbers on the meters in front of him, signalling, surely, a loss of engine power at just the worst moment.

The numbers dropped again. The third-stage engine was failing right in front of his eyes. If those figures limped any lower, the rocket simply wouldn't have the strength to shove the capsule into orbit and the launch would have to be aborted, with who knows what consequences for Gagarin at that extreme height.

But the numbers weren't falling now; they were holding steady. Low, but steady.

And now, very slowly, they were rising again, ticking upwards.

But were they rising enough?

It was a bad time to lose radio contact with the capsule. The numbers were starting to look good, but suddenly Gagarin could not be heard. On the ground, the silence was practically unbearable. Good numbers, but no contact – this was agony. Korolev, who had endured a sleepless night and had spent the morning taking pills for chest pains, was now white with fear.

But then, after a few seconds, contact returned. The voice of Gagarin crackled into the room. He was happy to report that he was in orbit.

He was almost babbling in his excitement, in fact.

'Everything is floating! Everything is floating! Beautiful. Interesting . . . I am watching a little star in the illuminator. It is going from left to right. The star has disappeared. It is disappearing, disappearing . . . I am watching the Earth, flying over the sea . . .'

He was seeing the curvature of the Earth from above, he was seeing the thin, translucent ring that binds the Earth's atmosphere, he was seeing space in its impossibly deep and oddly rich blackness. He was seeing things that nobody had ever seen before.

And none of us would ever see them in the same way again.

IV. DESPERATE TIMES

As he sat in the capsule of Mercury-Redstone 3 on 5 May 1961 and waited to become the first American in space, the launch pad escape options open to Alan Shepard included an explosive hatch release, a personal parachute which he wore strapped to his back – making the highly congested cockpit feel more congested still – and a set of external slide-wires, descending abruptly from the rocket and away. Those last were mostly for the evacuation of the ground crew in the event of a pre-launch emergency while they were still in attendance, but the astronauts seem to have enjoyed testing the system – effectively a fun-park zip-wire writ large.

Gunter Wendt recalls being present at a briefing session on emergency launch pad evacuation at Cape Canaveral with

Shepard, Gus Grissom, Pete Conrad and a pad safety officer, who was somewhat vexed when Shepard, without ceremony, right in the middle of the presentation, simply jumped on the wire and rode it down to the ground for fun.

'You can't do that!' the pad safety officer sternly insisted to Grissom, Conrad and Wendt on the platform.

'I'll go down and tell him,' said Grissom, and he grabbed the wire and followed Shepard down.

'You can't do that!' the pad safety officer repeated, more desperately, to Conrad and Wendt on the platform.

'I'll go down and tell him,' said Conrad, grabbing the wire and joining the others.

'They really can't do that!' the pad safety officer said despairingly to Wendt, appealing to his higher authority.

'I'll go down and tell them,' said Wendt, grabbing the wire and disappearing too.

The game of riding the wire for kicks came to an abrupt end one day when someone in the workforce parked a forklift truck at the bottom of the run and two staff members, not realising it was there, flung themselves down and narrowly avoided coming to a sticky end on its forks. At that point, Wendt felt he had no option but to turn teacher again. A strict rule was imposed: anybody using the wire for anything other than official business would be out on their ear and no longer working for NASA.*

Anyway, the wire was there if Shepard needed it, lying there 40 metres above Launch Pad 5 at the Cape that May morning. So far

---

* The sophistication of the evacuate-by-wire system increased over the years, becoming, in the Apollo era, a single-cab cable car that held nine and ran all the way to the perimeter fence, and, in the Shuttle era, a set of seven wires with baskets that held up to four people each.

so good, though. The countdown clock ticked down. Mercury-Redstone 3 was go for launch.

Inevitably, this flight was not quite the glorious bid for the history books that Shepard and NASA – and America as a whole, it sometimes seemed – had set their hearts on. Yuri Gagarin had seen to that, three weeks earlier. Wasting no time, the Soviet news agency had announced the arrival in orbit of a Russian cosmonaut while Gagarin was still in the air. If the landing had killed him . . . well, they would cross that PR bridge when they got to it.

So the Space Race was won – or, at least, its first heat. Actually its first two heats, because the Soviet Union hadn't just put a human in space, they had put a human *in orbit*, an achievement which was still fully two flights away in America's plans. NASA's press officer was awoken by a phone call in the early hours – a journalist wanting to get an official reaction to this story coming out of Moscow on the news-wires.

'We're all asleep down here,' NASA's man replied irritably, and hung up.

Cue the gift-wrapped headline: SOVIETS PUT MAN IN SPACE: SPOKESMAN SAYS U.S. ASLEEP.

Shepard, who was in training at Cape Canaveral, was roused that day in his Holiday Inn hotel room by a NASA public affairs officer who told him the news that he had dreaded hearing: that his dream of being the first human in space was over. Shepard turned on the television news channel to have it confirmed. He then allegedly thumped the table so hard in anger and frustration that the press officer thought he might have broken his hand.

He would have flown by now if NASA hadn't taken the cautious decision to insert two further test flights into the schedule – one of them, gallingly, featuring Ham the chimpanzee.* 'We had 'em,' Shepard is alleged to have said. 'We had 'em by the short hairs and we gave it away.'

Three weeks later Shepard had rallied. After all, the first American in space was still something to be, still a prize that all six of his Mercury colleagues would have snatched out of his bruised hand if he'd given them the merest hint that they could.

And America didn't seem deflated either. On the contrary, some half a million people had come out to the Cape to see the launch in person, and were now standing below Shepard, by the roads and in the dunes, across the beaches and on the roofs of cars and camper vans. Forty-five million more were watching the live broadcast on television, including a tense President Kennedy in the White House.

So Shepard found himself strapped in and facing upwards on the launch pad, primed, braced, the very focus of the nation's attention . . . and with a worryingly full bladder.

The suit for this mission had no urine-collection facility because, after all, why would Shepard need one? He was only going up in the air for just over fifteen minutes – a quick sub-orbital nip in and out of space. If he obeyed the ancient parental advice to go before he left home, he would be fine.†

* You didn't want to mention chimps around Al Shepard at the best of times. He was fed up with all the jokes, and once threw an ashtray at someone who joshed within earshot about needing somebody who was willing to work for bananas. Shepard told John Glenn he would like to have a 'chimp barbecue'. He wasn't kidding.

† Yuri Gagarin's suit, by contrast, did have a urination device, and the capsule carried a faecal waste unit. It was also loaded with ten days' worth of food and water, though that was longer than certain parts of the capsule's system, such as its dehumidifier, were designed to last, so it was an optimistic gesture to say the least.

Except these calculations failed to reckon with the possibility of delays on the pad . . .

After a breakfast of steak and eggs with orange juice, it was 5.15 a.m. when Shepard took his seat in the capsule – two hours before launch time. Three hours later he still hadn't launched and was now officially desperate to go – in both the traditional senses. Indeed, as far as Shepard's bladder was concerned, there could be no more holding on.

From the capsule came the request for a bathroom break.

Impossible, said Ground Control: there wasn't time to get Shepard in and out of the capsule.

'Then I'll have to do it in my suit,' said Shepard.

This suggestion caused consternation among the medical team. What if Shepard's urine pooled somewhere and shorted all their carefully positioned data-gathering sensors?

One solution: switch off the sensors, allow Shepard to urinate, grant a period of time for drying out, and then switch the power back on again. Fingers crossed, the sensors would survive.

After some debate, this was deemed to be the best way forward.

From Ground Control, and to his immense relief, Shepard heard the instruction: 'Do it in the suit.'

At 9.49, after nearly four and three-quarter hours of sitting in the spacecraft, the first American in space was finally airborne, if a little damp. He had been obliged to make himself, as Shepard put it later, 'a wetback'.*

In the violence of the launch, Shepard's voice shook. 'Roger,

---

* Accounts vary as to whether switching off the sensors succeeded in its objective. Some versions declare that a number of the sensors failed to work when the power was eventually returned to them; other versions state that none of them worked. It's not clear that Shepard was all that bothered either way.

lift-off, and the clock is started,' he could be heard saying through the judders. The vibrations in the craft were so extreme that he could barely read the instruments. But he picked out Florida and the Bahamas through his periscope viewfinder – albeit in black and white because he had inserted an anti-glare filter across the lens which he found he now couldn't remove. And he successfully tested the craft's manual controls, reorienting the capsule for re-entry, tilting the heat shield into the blast by hand.

Fifteen minutes and twenty-two seconds later, he was back, splashing down 302 miles from Cape Canaveral. He had flown to 116 miles above the Earth. And most important of all for the future of American space exploration, he had come back safely. In the White House, Kennedy breathed again.

The nation was ecstatic. Shepard was reborn as 'Shep' in a hundred triumphant headlines. He was given a parade in Washington and a reception on the White House balcony, where, in a suit and tie, he stood with his hands behind his back and his chin held high and received a NASA Distinguished Service Medal, which, somewhere between its hinged box and Shepard's lapel, President Kennedy somehow managed to drop unceremoniously on the floor. ('This decoration, which has gone from the ground up . . .' Kennedy smoothly ad-libbed as he finally handed it over.)

As the engineer Werner von Braun said, 'The future of our entire manned space programme hinged on the success of this flight.' And success was duly delivered. Now all systems were go. Ahead lay a long succession of escalating mission triumphs during which Shepard himself would become the sole Mercury astronaut to walk on the Moon.

And ahead lay a period when spaceflight and American astronauts could be relied upon to provide an exultant demonstration of American infallibility almost whenever one was needed – a period that would last a long time.

But this was spaceflight. It couldn't last forever.

V. FROM A CLEAR BLUE SKY

On 27 August 1984, President Ronald Reagan announced NASA's Teacher in Space Project – a plan to fly a teacher as part of a Space Shuttle crew in 1986. 'When the Shuttle lifts off, all of America will be reminded of the crucial role teachers and education play in the life of our nation,' Reagan declared. 'I can't think of a better lesson for our children and our country.'

Eleven thousand teachers applied for the role. Over the course of the next year, those applicants were whittled down to ten finalists who were tested and evaluated at the Johnson Space Center, and the winners were announced by Vice-President George H. W. Bush in July 1985. The prize role of primary candidate went to Christa McAuliffe, a history and social studies teacher at Concord High School in New Hampshire, with Barbara Morgan, a science and English teacher at McCall-Donnelly Elementary School in McCall, Idaho, as back-up.

McAuliffe was granted a year of leave from school to train, at the end of which she was appointed to the six-person crew of Space Shuttle *Challenger* on Mission STS-51L. The mission would be carrying three experiments designed by students: one studying chicken embryo development in space, one considering the effects of weightlessness on grain formation and strength in metals, and one looking at how a semi-permeable membrane could be used to direct crystal growth. On the sixth day of the mission, via satellite and the Public Broadcasting Service, McAuliffe would teach two lessons from space, one just before the lunch recess and one after it. McAuliffe described it as 'the ultimate field trip'.

When McAuliffe, smiling broadly, boarded *Challenger* on the sunny but bitterly cold morning of 28 January 1986, she did so in the company of a talented and storied crew. STS-51L's Mission Specialists were Judith Resnik, who flew the maiden flight of *Discovery* in 1984 and became the second American woman in space; Ellison Onizuka, who was on his second Shuttle mission, having previously become the first Asian American in space; and Ronald McNair, from the 1978 intake, the second African American in space, flying his second Shuttle mission. STS-51L's Payload Specialist was Gregory Jarvis, a scientist who had been in the astronaut programme for two years. The mission was commanded by Francis 'Dick' Scobee, who flew the fifth *Challenger* flight, in 1984, which successfully repaired the malfunctioning Solar Maximum satellite, thereby importantly establishing the Shuttle's capacity as a tool for satellite maintenance. The pilot, meanwhile, was Michael J. Smith, a former Navy test pilot, flying his first mission.

Naturally, the launch was the subject of avid interest in American schools, interest stoked by a massive outreach programme in the build-up. It was early morning on the west coast when the crew boarded, so pupils were still mostly making their way in, but schoolchildren to the east were primed and ready in front of school television screens, not least at Concord High, where excitement was especially vigorous. A small group of McAuliffe's students had been taken to Cape Canaveral to witness the launch. Others, dressed in paper party hats, watched from the school cafeteria.

They saw *Challenger* leave the pad against a perfectly blue sky. For all that it had been repeated many times by now, the sight of a Space Shuttle climbing aloft, clasped to the almost temple-like architecture of its orange external tank, was still an unfailingly awe-inducing sight.

The launch proceeded entirely smoothly for more than a minute. Around seventy seconds into the flight, with the Shuttle approaching 46,000 feet, Ground Control heard Dick Scobee say, 'Go at throttle up.'

After that, there was only one more entry on the Shuttle's voice recorder: the voice of Mike Smith saying, 'Uh-oh.'

One second later, the spacecraft blew apart.

The crew capsule, propelled higher in the explosion, endured a 60,000-foot fall, hitting the water at 250mph. It came to rest 85 feet below the ocean's surface, some 15 miles east of Cape Canaveral. Navy divers took six weeks to find it. Someone described it as resembling a scrunched-up ball of tin foil. Debris from the *Challenger* would continue to wash up on Florida beaches for more than a decade.

According to the independent Rogers Commission, under former Attorney-General William P. Rogers, the technical cause of the disaster was a fault with an O-ring which had failed in the sharp weather – one of fourteen O-ring near misses in Shuttle history that the report brought to light. It emerged that NASA engineers had argued that the weather was going to be too cold for the launch on that January morning – unprecedentedly cold for a launch day, in fact – and had recommended postponing it. But NASA appears to have been preoccupied with flight targets – with reaching twenty-four launches per year and thereby justifying the Shuttle's existence as America's sole space-freight and personnel carrier – and the recommendation was ignored. *Challenger* launched, and *Challenger* exploded.

Here was another graphic illustration of the eternal rupture between generals and troops, bosses and workers, administrators and practitioners – between the people making the judgement calls on acceptable risk and the people actually facing that risk: the people strapped into their seats on the tops of the rockets, the people still on the launch pad when everyone else had retreated to a safe distance.

The Space Shuttle project never recovered from the loss of *Challenger*. Its scope was narrowed, its blithe certainty about itself gone and irrecoverable. The destruction of *Columbia* and its crew in 2003 (which we will come on to later) brought about the programme's end. But *Columbia* was a tragedy that occurred on re-entry, and was therefore more distant, less intensely monitored in real time and to that extent less raw. The visceral awfulness of *Challenger*, and the shock value it retains for all of us who are old enough to remember it, was that it happened during launch, in plain sight – and, more than that, in front of, in effect, a specially invited audience of schoolchildren stretching all the way round the world.

These were risks that had been known to astronauts all along. These were realities that astronauts negotiated and variously reached an accommodation with in the course of their training and their working lives. But now they were everybody's to think about, including children.

In the wake of the disaster, it was reported that children dreamed of explosions, fires, injury and death. This was not what people meant when they spoke of spaceflight encouraging the young to dream. Yet at the same time these things were a part of spaceflight and how could you deny that? A study of children's reactions to the *Challenger* disaster published in the *American Journal of Psychiatry* in 1999 found that most symptoms of trauma 'dramatically faded' over a fourteen-month period, but that 'adolescents' diminished expectations for the future increased'. One boy told the study: 'I've been worrying a lot – what it feels like not to exist.' He was eight.

This was among the lessons of *Challenger*: that there were risks involved in bringing everyone along for the ride. Yet this was spaceflight; its risks to the humans at its centre could not be wished away. Indeed, spaceflight's relationship with risk was integral to the greatness of its endeavour, and you could even say part of its inspirational force. At its heart were people who had decided that exploration beyond Earth's boundaries was valuable enough to be worth the manifest dangers to themselves, and the boldness and clear-sightedness of that decision was at the core of our respect for those people every time they climbed into rockets.

But those risks could be mitigated. And with *Challenger*, NASA had clearly fallen short in its duty to mitigate them.

A long period of sober reflection and corporate self-examination followed. After STS-51L, no more civilians flew on the Shuttle for over twenty years – until 2007, in fact, when, with a symmetry that felt restorative, Barbara Morgan, McAuliffe's reserve, having gone back into training, flew as a Mission Specialist and robot-arm operator on STS-118, an assembly mission to the ISS. Morgan then returned to teaching, at Boise State University. NASA had hoped to fly Shuttle orbiters until at least 2020, but in 2010, following the recommendations of the report into the *Columbia* disaster, the Shuttle was decommissioned and the programme closed down.

One thing that survived intact in the *Challenger* wreckage was a football. It had been packed for the journey by Ellison Onizuka, and was the ball used by his daughter's high school team. It was returned to her. She in turn donated it to a *Challenger* memorial display at her school, Clear Lake High. Thirty years later, the NASA astronaut Shane Kimbrough, whose own daughter was a student at Clear Lake, asked if the ball could accompany him on his mission to the ISS. He took a photograph of it floating against the backdrop of the Cupola observatory module window – a poignant and somehow defiant image: Onizuka's dropped ball, picked up and run with.

Even before it ended in disaster, Christa McAuliffe's participation in STS-51L had begun a debate about the form that Shuttle crews should take. Eight years of cultural change had occurred since the 1978 NASA astronaut draft irrevocably broadened the corps beyond its military-test-pilots-only origins. But how wide open, in fact, could and should the door be for what was now euphemistically referred to as 'spaceflight participation'?

Astronauts in particular, unsurprisingly, tended to have strong views on this. Going into space with people whom you know to have trained as hard and as long as you have, or even harder and longer, brings with it certain reassurances and guarantees. Going into space with someone who has had a relatively short preparation period and whose character under pressure you know little about beyond the word of your space agency . . . that's bound to feel different.

The PR and media spin also irked some of the astronauts. The plain fact was, the primary objective of STS-51L was the launch of a multi-million-dollar satellite system for NASA and the US Air Force, and the transport into orbit of an instrument designed to observe Halley's comet. It was not to put a teacher in space – although, of course, that was the way the mission became characterised.

It didn't help when a twenty-four-hour launch delay to STS-51L meant that, within the flight plan, McAuliffe's live lesson from space would be going out on a Saturday, with no students at school. There wouldn't be much point in that, clearly, so the launch plan was rewritten and McAuliffe's lesson moved.

This change caused some bewilderment among the astronaut

corps. Aviators have it drummed into them over and over again: 'Plan the flight and fly the plan.' A last-minute change to the flight plan to accommodate what was, from the astronaut's point of view, a secondary mission? Such a thing was not only entirely unprecedented, it was anathema as far as many in the corps were concerned.

But here was a problem which the nature of the Shuttle itself had partly foreshadowed. Hierarchy was embedded in the very structure of the spacecraft. Just like the passenger ships of old it had two decks and accordingly, by implication, two classes of crew: a flight deck with windows where the Shuttle got flown; and a mid-deck, windowless, where the 'lower orders' sat – who might, incidentally, be engaging in extra-vehicular activities (EVAs) or mending the Hubble telescope or doing any number of other highly specialist technical tasks but who nevertheless were not felt to be holding a ticket as valuable as those of the pilot and the Commander upstairs.

From such assumptions grew very easily the notion that the Shuttle could carry passengers. Was there not room for them downstairs? NASA's management did not endear itself to the astronaut corps when it began handing Shuttle seats to politicians for what was called the Politician in Space programme. This all made obvious sense from the agency's point of view. It was a great way to increase profile and curry favour in Washington, without which NASA could hardly survive. But it grated on the trained astronauts, not least those still waiting for a mission. The queue for assignments was already long – and now politicians were being jumped into it?

It grated especially painfully when Congressman Bill Nelson, now NASA's Administrator but then a Democrat Congressman in the House of Representatives, was given a seat on Space Shuttle *Columbia* at the beginning of 1986, bumping Gregory Jarvis, one of NASA's own, onto the next flight. The next flight was *Challenger* STS-51L.

But it wasn't only politicians. There were strong rumours in the 1980s that the country singer John Denver would be given a slot on the Shuttle. Moreover, as well as appointing McAuliffe for STS-51L, NASA approached the producers of the educational children's show *Sesame Street* about getting a character from the show, Big Bird, and the actor who played him, Caroll Spinney, to participate in that *Challenger* mission – a move presumably calculated to extend still further the flight's reach to children. NASA confirmed in 2015 that it had negotiated with the programme but 'the plan was never approved'.

Still, whoever they were, if the Shuttle could carry passengers, then the implication was very strongly that the orbiter was just some kind of glorified commercial 747, not an enormously dangerous spacecraft that, incidentally, offered its crews no emergency evacuation system, barring those wires and baskets outside on the launch pad. (Ejector seats featured in the non-orbital test vehicle, the *Enterprise*, but in no other iteration.)

And just to be clear, the attitude of astronauts to the notion of Shuttle passengers was not simply an instinctive resistance to people beyond the astronaut tribe. It extended to John Glenn. In 1998, the first American to orbit the Earth and one of the certified heroes of US aviation returned to space on board Space Shuttle *Discovery* as part of the crew for STS-95. He was seventy-seven. At the time, that made Glenn the oldest person ever to fly in space, and his participation was billed as an opportunity to study the effects of spaceflight on the ageing. Many astronauts, though, regarded that scientific purpose as the thinnest of veneers. Also, wasn't there still a queue for these precious and hard-won seats? Glenn was a US Senator who came into the mission in a blaze of publicity, and for a certain sector of the astronaut corps this was politics and showbusiness and Glenn should not have been up there.

The *Challenger* STS-51L crew never reached the Karman Line. Accordingly Mike Smith, fully trained but a rookie, did not die as a flown astronaut by the official definition. Mike Mullane, among others, forcefully argued that NASA should adjust its terms in the wake of that accident. After all, it was as true for Smith as it was for everyone on that flight: he had climbed into that Shuttle and readied himself for the launch, knowing full well the stakes.

And yet still he had climbed in.

As soon as the hold-down bolts blow under the rocket, Mullane suggested, you're a flown astronaut.

I wouldn't argue with him.

Also, as someone who was a thirteen-year-old schoolboy when *Challenger* happened and who saw the news footage and shared the deep-seated shock of it with his friends and talked about it with his teachers and his parents, I find myself completely in sympathy with the eighteen-year-old respondent to that *Journal of Psychiatry* study who said: 'When the sky is that certain blue of the day of the launch, I always think of *Challenger*.'

'But,' he added, 'you always recover. You move on.'

As an astronaut, I believe in doing that too.

CHAPTER FIVE

GETTING THE JOB DONE

'Every day is a good day when you're floating.'

– Anne McClain, NASA Artemis astronaut

I. FLYING BLIND

By the beginning of 1962, NASA were ready to make John Glenn the first American to orbit the Earth. The question was, what were they going to get him to do while he was up there? Ideas for trials and experiments that could be conducted during the flight of Mercury-Atlas 6 were floated in a steady stream as the scientists working with NASA tried to wring the most out of this unprecedented testing opportunity.

By this time, the Soviet Union had flown two orbital flights – Gagarin's in April 1961, and then Gherman Titov's in August. Aboard Vostok 2, Titov had spent an entire day in space, completing seventeen orbits. He had become the first human to sleep in space and the first to film the Earth, returning with ten minutes of footage on a movie camera. He had also been copiously sick, although this was not reported at the time. The Soviet Union weren't sharing the details of their findings from these flights, beyond the fact that both cosmonauts had come back alive. In terms of what prolonged spaceflight actually did to the human body, both during and afterwards – and what this implied for future missions – it was still all for America to discover.

Glenn thus found himself in two ostensibly conflicting roles: part fearless aviator and boundary-busting pioneer, part sitting target and space agency guinea pig. As the years unfolded, that duality would come to sit at the heart of what it was to be an astronaut and to do an astronaut's work.

At one point during the planning meetings, conversation turned to the possibility of harvesting some data regarding human orientation in weightlessness. Someone wondered whether, at a set point in the flight, Glenn might reach for certain switches on the cockpit's console with his eyes open, and then close his eyes and try to reach for those switches again to discover whether he could. The suggestion was that this simple 'before and after' test might yield valuable lessons about the persistence of spatial awareness and small motor skills in the absence of gravity.

Did a ruminative silence descend on the room while everyone absorbed this proposal? It's nice to imagine one. At any rate, it seems to have been the astronaut Gordon Cooper who pointed out bluntly that, in the name of science or otherwise, fumbling about blindly for switches would be a pretty stupid thing to do inside a spacecraft making its first orbiting flight. The blind switch-finding orientation experiment wasn't mentioned again.*

NASA would find more straightforward ways to monitor Glenn's faculties, including his eyesight. He travelled to space with a miniature version of the letter chart you might typically

---

* The effects of microgravity on spatial awareness continue to intrigue scientists and have been the focus of several studies and scholarly articles since. I can't pretend to be entirely on top of the latest findings but I can report that the only advice in this area offered to me before my mission was to watch my head flying through the ISS's hatches. And I can further report that not once did I crack myself on one, despite flying through several at speed.

see hanging in an optician's. It was fixed at the top of the instrument panel, directly in Glenn's eyeline. Every twenty minutes throughout the nearly five-hour flight he had to read the smallest line on the chart that he could manage, and a record was kept. He also seems to have taken part in an experiment investigating the effects of spaceflight on human digestion which required him periodically to take xylitol tablets.*

For his own part, Glenn wanted to take photographs during his mission, but flight directors thought this would be too much of a distraction. He talked them into it, though, and took up his own 35mm Minolta camera, its controls slightly modified for operation with thick gloves on. (Even then, loading film proved fiddly, and in the course of his fumbling Glenn let a film canister get away from him. It was later found behind the instrument panel.)

As well as thinking about what he might do during the flight, Glenn himself seems to have been concerned about what might happen at the end of it, when he landed. In particular he raised the possibility of his spacecraft coming off course and arriving among a remote people – people with no access to news outlets, say. What if people not up to speed with developments in the Space Race reacted adversely to the – let's face it – potentially startling sight of someone in a silver suit descending from the sky in a charred tin box?

As we saw earlier, in common with all the Mercury astronauts Glenn had undergone extensive survival training by this point, in both desert and jungle terrains. But apart from the occasional encounter with predatory wildlife, none of those exercises had

* Xylitol is a natural sugar alcohol, often used as a sweetener in food. The digestion experiment didn't seem to require Glenn to eat much himself. The only thing he consumed on his flight was a solitary tube of apple sauce. He had, though, breakfasted heartily on steak and eggs in the NASA astronaut tradition.

placed the astronauts among actively hostile populations and forced them to improvise diplomacy. (Not that diplomacy had ever been anything that Glenn, in particular, seemed likely to struggle with.)

NASA were able to guarantee Glenn that, wherever he landed, they were confident of getting US forces to him within seventy-two hours. But, as Glenn pointed out, seventy-two hours was quite a long time, especially if you happened to be spending it among people who considered you to be a threatening alien invader.

So it was agreed that Glenn would carry with him a card, bearing a message translated into seven languages and written out phonetically for easier reading. Come the moment, Glenn could whip out the card and, with luck, broaden his chances of explaining himself to anyone puzzled or made defensively anxious by him.

The message eventually drafted was, in its English version: 'I am a stranger. I come in peace. Take me to your leader, and there will be a massive reward for you in eternity.'

How an American astronaut, even one as capable as John Glenn, came to be in a position to offer rewards in eternity was unclear, and it would be interesting to know how he would have responded had he ever been called upon to uphold his half of that particular deal. It's also possibly not just cynical to wonder whether giving Glenn the wherewithal to offer a reward in the actual here and now might have granted him a stronger bartering position in the circumstances.

Still, a catchphrase was born. Over the ensuing decades, the command 'Take me to your leader' was an expression that sank deep into English-speaking culture, a gift that kept on giving for screenwriters, cartoonists and comedians.

Originally scheduled for January 1962, Glenn's launch was

scrubbed multiple times and he didn't get into the air until late February. Phoning his wife Annie at home just before the flight, Glenn said, 'I'm just going down to the corner store to get a pack of gum.'

Annie Glenn replied, 'Don't take too long.'

It was the same ritual exchange they had been having before Glenn's missions since he became a military pilot, to ease Annie's fear.

He saw a dust storm as he flew over Africa. The city of Perth in Australia switched on its lights for him as he passed over, and he saw those too. He also saw what became known as 'John Glenn's fireflies' – frozen particles of gas from the thrusters, which flew along with him for a while.

Meanwhile, down below, the NASA scientists reaped their precious data on their astronaut's heart-rate, his eyesight and his digestive system.

To what extent was Glenn an observer, and to what extent was he the observed? To what degree was he experimenting, and to what degree was he the experiment? In what measure was he the magnificent agent of his own destiny, and in what measure was he just a finely adjusted cog in a complex scientific and political machine?

Every age of spaceflight has sought to define astronaut work by finding answers to those questions. And in every age the answers have been different.

II. A BAD DAY AT THE OFFICE

On a Wednesday morning in March 1966, NASA launched Gemini 8, the sixth crewed flight in the Gemini programme. The astronauts on board were both rookies. In the pilot's seat was David

Scott, a former US Air Force test pilot, aged thirty-three. In the Commander's seat was thirty-five-year-old Neil Armstrong.

The mission had, typically by now, a significant science content. During their three days in space, Armstrong and Scott were to oversee experiments in the growth of frogs' eggs and in nuclear emulsion, to observe atmospheric cloud formation and to conduct various types of photography, both of space and the Earth's terrain.

But the most important experiment, or certainly the most complex and dangerous one, would be performed with the spacecraft itself. Once in space, Armstrong and Scott would chase after and rendezvous with an empty Agena module, launched into orbit ahead of them as a target. Their task was to catch up with the Agena and dock with it. This was going to be the most complex mechanical manoeuvre in space that NASA had ever attempted, and a trick that not even the Soviet Union had yet pulled off. Geminis 6 and 7 had laid the ground for rendezvous between two spacecraft, with Wally Schirra and Tom Stafford bringing their capsule to within a few feet of Frank Borman's and Jim Lovell's. Now Gemini 8 would go the whole way and lock two craft together – at 161 miles above the Earth and while travelling at 17,500mph.

The first phases of this precarious tango in space passed off so easily that it seems to have surprised even Armstrong. The Gemini flew smoothly, tracking the Agena target by radar, flying in close and then 'station keeping' – travelling in tandem with it. Eventually, after circling the Earth alongside the Agena, Armstrong was ready to move in for the docking.

That operation, too, was a flawless success. An enthused Armstrong called down: 'Flight, we are docked! Yes, it's really a smoothie!'

The capsule went out of communication range almost immediately after that, so Ground Control didn't hear what happened next. No sooner were the two craft together than they began bending apart at the join. Then, locked in their crooked embrace, they began to roll and yaw – moving from side to side. Then the roll turned into a spin. The worry was that, in all this movement, the docking adapter wouldn't be able to take the strain and would simply snap.

Here was an extremely rare instance of an emergency for which there had been no rehearsal in the simulators. Assuming that the abrupt loss of control was down to a fault with the Agena, Armstrong decided the best course was to separate the two spacecraft again. Using the thrusters in the Gemini's Orbit Attitude and Maneuvering System (OAMS), he managed to slow the movement to a point where he was happy to undock. Then, giving the Gemini an extra kick of power to ensure that it didn't collide with the Agena on the way out, he pulled back.

But it was when they were separated from the Agena that the problems really began. Now, out on its own, the Gemini started to somersault.

When radio communication resumed, Mission Control heard the juddering voice of a clearly stressed Scott say: 'We have serious problems here. We're . . . we're tumbling end over end up here.'

The problem wasn't with the Agena at all, it now turned out; it was with the Gemini they were sitting in. Unbeknown to Armstrong, during the docking a short circuit had caused one of the craft's eight OAMS thrusters to stick open. He was now using those same thrusters to try to bring the Gemini back under control, and yet – unsurprisingly, with one of those thrusters permanently pulling at the vehicle – nothing was working.

In fact, the predicament was only worsening. The capsule was now spinning through 360 degrees every second, and pitching and yawing at the same time – like a theme park ride from hell. How much of this violent motion could the Gemini take before bits started to break off? More critically, how much of this violent motion could Armstrong and Scott take? Spun and tumbled in this spacecraft-turned-centrifuge, it was hard for them to cling on to their senses. Both astronauts found their vision beginning to blur. Both were in danger of passing out. Only by clenching his neck muscles to keep his head stiffly in place could Armstrong hold the spacecraft's controls in something like focus.

Somehow, in the middle of this radical disorientation, Armstrong retained the presence of mind to assess his options. Despite everything, he managed to switch off the OAMS, the main source for manual control of the craft which had been of no help to him so far, and instead powered up the re-entry control thrusters.

Situated at the front of the Gemini, the re-entry thrusters were only intended to be deployed on the capsule's return. Armstrong knew that, by using them now, he would be burning through fuel that was reserved for the critical last phase of the journey – and who knew how much that might complicate things later?

On the other hand, if he couldn't get the capsule under control, there would be no last phase of the journey anyway.

It took thirty seconds – thirty seconds during which Armstrong and Scott were being rolled around like clothing in a tumble dryer while Ground Control, helpless down below, began to fear the worst. Yet with repeated squirts of power from the re-entry thrusters in just the right quantities, at just the right moments, Armstrong managed to get the capsule stable again.

The Gemini was back under control and Armstrong and Scott could gather themselves. But the mission was now officially aborted. Armstrong set the capsule for re-entry – for which there was, blessedly, still enough fuel in the tanks – and, after blazing back into the atmosphere, he and Scott splashed down some way off course, in the western Pacific, 600 miles south of Japan.

There they bobbed and thrashed in the waves for forty minutes until rescue planes arrived, dropping divers who threw a flotation collar over the capsule to ensure it didn't sink. As if their journey hadn't already involved enough pitching and yawing, Armstrong and Scott then sat, rolling on the sea, for a further two hours until a US Navy destroyer came to collect them. The immediate reward for surviving NASA's closest brush so far with death in space was a grim bout of seasickness for both of them.

It's clear that Armstrong and Scott were only saved that day by Armstrong's prodigious piloting skills and presence of mind. Yet his mentality as an astronaut seems to have precluded him from taking much pleasure in what had happened, let alone banking any credit for it. He was apparently downcast in the immediate wake of this adventure, the satisfaction of having survived against the odds not figuring as large in his mind as the dissatisfaction of failing to complete the mission. As far as Armstrong was concerned, he had been obliged to fly home early, therefore he had come up short.

So, if we choose to consider Armstrong's flying on that occasion as the work of an aviation genius at the top of his game, we must also acknowledge that this is not how Armstrong himself felt about it. Just as he would be in the wake of the accident with the 'Flying Bedstead' Lunar Module simulator just over two years later, he was the model of understatement about the rare drama of his first time in space.

'It was one of those bad days,' he said afterwards.

He just had to hope there would be better ones.

III. CORNED BEEF ON RYE, HOLD THE MUSTARD

It was clear from the beginning that astronaut work was dangerous and complex and that the consequences of even slight mistakes when it mattered were likely to be dire. Was there any room for mucking about in such a workplace? The first astronauts clearly thought so. But the authorities didn't always agree. The grave incident of the smuggled sandwich, and everything that followed from it, is a case in point . . .

During his pioneering orbit of the Earth in 1961, Yuri Gagarin consumed two tubes of meat paste and one tube of chocolate sauce – not exactly a feast. But then, launching from the Baikonur Cosmodrome on the desert steppe of south Kazakhstan, Gagarin didn't have pre-flight access to Wolfie's Restaurant Sandwich Shop & Cocktail Lounge at Cocoa Beach.

Wolfie's was a franchise joint inside the Ramada Inn that announced itself loudly to the highway with an asymmetrical sign hoisted high on a V-shaped mount and the Wolfie's logo: a cartoon of a wolf in yellow overalls somehow looking both hungry and sated at the same time. For the NASA astronauts of the sixties, working at Cape Canaveral and frequently staying nearby in the Cocoa Beach Holiday Inn, Wolfie's was a regular late-night hang-out.

The place was run by Carl and Pat Ransom. One night in 1965 Wally Schirra was sitting at the bar, talking to Carl about how bad the food was that astronauts were given to eat in space – tasteless, pulpy stuff in tubes and sachets. Carl remarked that it was a pity NASA hadn't asked him and Pat to do the catering.

Which gave Schirra an idea.

Gemini 3, the first crewed Gemini mission, flown by Gus Grissom and John Young, was due to launch from the Cape very shortly, on 23 March. Maybe Wolfie's could cater it . . . but secretly.

It was Schirra, then, who bought the sandwich – corned beef on rye, no mustard, no pickle. And it was Schirra who helped John Young smuggle that sandwich, wrapped tightly in cellophane, into a pocket in Young's spacesuit on the morning of the flight.

At last! Something decent to eat on a space mission. OK, so, what with everything, the sandwich was already two days old by the time it reached Young's pocket. But even so.

This was all a prank, of course, intended first and foremost to get a reaction from Gus Grissom, who would be next to Young in the spacecraft when, out of nowhere, he pulled out his non-regulation packed lunch. But it was also a joke that contained a barely concealed dig at NASA. The feelings about space catering that Schirra had expressed at the bar in Wolfie's were widely shared among his colleagues: the food didn't just *need* to be sucked, it *actually* sucked.

The smuggling of the sandwich went without a hitch. Young then waited for his moment. One hour and fifty-two minutes into the flight, and with Gemini 3 safely in orbit, Mission Control heard the following dialogue take place in the capsule:

Grissom: What is it?

Young: Corned beef sandwich.

Grissom: Where did that come from?

Young: I brought it with me. Let's see how it tastes. Smells, doesn't it?

Grissom: Yes, it's breaking up. I'm going to stick it in my pocket.

Young: Is it? [A pause.] It was a thought, anyway.

Grissom: Yep.

Young: Not a very good one.

Grissom: Pretty good, though, if it would just hold together.

Young: Want some chicken leg?

Grissom: No, you can handle that.

The chicken leg wasn't another joke on Young's part. Along with a hot dog, it was one of a number of dehydrated 'regular food' packages that had been stowed on the capsule for the astronauts to trial – a sign, as it happened, that NASA were already listening to the astronauts' feedback about the food and trying to make it more palatable, or at least more recognisable.

Yet even at two days old and smelling a bit, the corned beef sandwich seems to have looked more tempting than the dehydrated chicken leg.

Sadly, though, Grissom wasn't exaggerating when he said the sandwich was breaking up. By the time he had transferred the rogue item to his pocket, a small cloud of breadcrumbs, caraway seeds and flakes of corned beef had formed in the capsule. Grissom and Young gathered up the debris as best they could, and went back to eating the supplied food.

In its entirety, the incident had absorbed fewer than thirty seconds of a four-hour-and-fifty-two-minute flight, including the clean-up. Yet it was set to absorb an awful lot more. Little did Young and Grissom suspect that the corned beef sandwich they had just unwrapped would cause not just a moment of remorseful reflection at NASA, as it was intended to, but a Congressional inquiry in Washington.

And no, the inquiry was not launched in righteous indignation at how poorly astronauts were being fed. On the contrary, it took the form of a dressing-down for all involved in the scandal that people would doubtless have been calling 'Sandwichgate' but for the fact that Watergate hadn't yet happened.*

When the story of the uncommissioned snack got out, it caused little amusement among members of the House of Representatives' Appropriations Committee. Indeed, the feeling was that larking around with sandwiches was not something the country's publicly funded astronauts were being paid to do, particularly when one of

* Changing the labelling of scandals by journalists forever more, the exposure of the Nixon administration's involvement in a burglary at the Watergate office building, headquarters of the Democratic National Committee, occurred seven years after Gemini 3 flew, in 1972. The Watergate scandal led directly to Nixon's resignation in August 1974.

the prime projects of the Gemini mission just completed was to test the emulsified astronaut food and various cubes carefully coated in gelatin that NASA had spent good time, money and energy developing.

There were questions to answer about this sandwich, then – and not just why Schirra hadn't asked for either mustard or pickle. James Webb, the NASA Administrator, George Mueller, the Associate Administrator for Manned Spaceflight, and Bob Gilruth, Director of the Manned Spacecraft Center, were summoned to Washington to explain themselves.

Representative George Shipley of Illinois came at them as follows: 'My thought is that . . . to have one of the astronauts slip a sandwich aboard the vehicle, frankly, is just a little bit disgusting.'

He may have had a point, not least given the age of the sandwich.

The grilling – if we can put it that way – of NASA's top brass went on in this fashion for a whole morning. Issued at the end of the session, Mueller's humble appeasement of the committee would go down in legend. 'We have taken steps,' the NASA executive solemnly assured the assembled Congressmen, 'to prevent recurrence of corned beef sandwiches in future flights.'

The message to the astronaut corps back in Houston and on the Cape was twofold. First, that astronaut work was serious. Highly serious. Missions were no time to be fooling around, not with sandwiches or with anything else. Second, that corned beef sandwiches were off the menu.

Except, of course, at Wolfie's. Outside the Ramada Inn, the restaurant gained a proud new sign.

'Try our corned beef,' it said. 'It is out of this world.'*

---

* The (allegedly) original space-bound sandwich is preserved in acrylic and displayed at the Grissom Memorial Museum in Mitchell, Indiana.

IV. HEAD COLDS AND 'MICKEY MOUSE' TASKS

Of course, fooling around with sandwiches in the workplace was one thing; directly refusing orders from the ground was something else altogether.

Apollo 7 launched in October 1968. On board were Wally Schirra, Donn Eisele and Walter Cunningham. Schirra, of course, was the vastly experienced Mercury Seven draftee whose cool thinking aboard Gemini 6A we witnessed in the previous chapter. Eisele and Cunningham were rookies but had been training as NASA astronauts for five years. This was a strong crew for a vital flight.

Their mission was to test the reliability and habitability of the Apollo's Command and Service Module and show it to be fit for purpose for a Moon shot later in the year. They were also to attempt the first television broadcasts from space, a publicity opportunity that NASA were particularly excited about but which Schirra – sounding, perhaps, an early alarm bell about his own mood ahead of this trip – had described dismissively as 'a dog-and-pony show'.

For the astronauts, the mission would prove to be both a complete success and an outright disaster, depending on how you looked at it. In the summary of Eugene Cernan, who was a member of Apollo 7's back-up crew, 'For 11 days [Schirra, Eisele and Cunningham] did two things extraordinarily well – successfully perform every test assigned and piss off about everybody in the program, from grunt engineer to the flight director.'

Context is especially important here. Apollo 7 was the first crewed flight since the fire on the pad that killed the Apollo 1 crew a year and nine months before, and feelings and anxieties around the mission were inevitably heightened at all levels.*

---

* After Apollo 1, there was no actual Apollo 2 or 3. Instead a series of tests took place with a different numbering system. Apollos 4, 5 and 6 were non-

Schirra, Eisele and Cunningham had lost three colleagues in that accident, and Schirra had lost one of his closest friends, Gus Grissom. Those two shared the bond of being the only astronauts to have been involved since the very beginning of crewed spaceflight and to have worked on all three of its initial programmes, Mercury, Gemini and Apollo. Schirra felt Grissom's loss keenly. He had already taken the decision, at forty-five, to leave NASA after the completion of the Apollo 7 mission.

The accident felt more than usually close to Eisele, too: he was originally selected for the Apollo 1 crew but a shoulder injury forced him to stand down; Roger Chaffee had replaced him. The fact that his own good luck had been Chaffee's dire misfortune would unavoidably have haunted him.

On top of that, America's declared plan to have an astronaut on the Moon by the end of the decade – a wildly optimistic target from the moment it was set – was now running late and the burden of meeting that deadline lay heavily on the successful operation of Apollo 7. The pressure was well and truly on.

During the mission's production phase, Schirra, as Commander, policed the construction of the spacecraft tightly. The in-depth examination of NASA's modes of operating which had inevitably followed the Apollo 1 disaster had definitely led to increased powers of input for the astronauts – the people in the capsules when things went wrong. Yet sometimes, if Schirra saw something he didn't like about the Command Module that was being assembled for his mission, he would pull rank on the engineers with the line 'You don't have to fly it – I do.' Which was obviously true and unanswerable, but nobody in his position had made the point

---

crewed tests of the nearly complete system with actual modules of the Apollo spacecraft as it was intended to be flown.

quite so aggressively before, and such withering dressing-downs did little to endear Schirra to the engineers.

That brittle haughtiness accompanied the mission's Commander onto the flight and, in the way of these things, it seems to have filtered outwards to influence the behaviour of his fellow crew members. In Cernan's phrase, they 'bitched and growled' throughout the mission – all of that bitching and growling, of course, occurring publicly over the communication lines with Mission Control.

They complained about the food, of course, which was only traditional. But they also complained about the fans being too noisy. They complained about the sleep rota, which required one of them to be awake and working at all times, meaning that the capsule never fell entirely quiet. And they complained about what they perceived to be the mundanity of some of the tasks they were set – 'Mickey Mouse jobs', as they referred to them. Most of these – not unreasonably on what was essentially a critical shakedown flight – involved repeated checks of the spacecraft's equipment.

'I wish you would find out the idiot's name who thought up this test,' Schirra announced at one particular low point.

Eisele joined in, bringing up another task on their checklist that was irking them. 'That is a beauty also,' he said, sarcastically.

Again, they were by no means the first astronauts to grumble about the seemingly trite elements of some of their work, and nor would they be the last. But they were certainly the first to do so mid-mission, at a time when everyone could hear them, including the actual devisers of some of those trite elements.

Around fifteen hours into the flight all three of the crew developed head colds, with runny noses, which are especially unpleasant in zero gravity and which required them all to take aspirin and use copious quantities of decongestant.* This did nothing to improve the mood in the confined capsule. In the words of Cunningham, 'It turned our cosy little spacecraft into a used Kleenex container.'

Meanwhile, some aspects of the spacecraft were a bit under the weather too. Condensation on the coolant lines in the cockpit caused water to cling to all the surfaces. The crew had to rig up a urine drain hose and pump the water out into space. Three of the five windows were constantly misted over, affecting the astronauts' ability to perform some of the observational tasks. None of this in any way diminished their irritability, which wasn't exclusively reserved for the workers on the ground: they turned it on each other, too. Eisele and Schirra had a heated dispute over a navigational issue, all caught on a live microphone, and the pair of them appeared to sulk for several hours afterwards.

And then there were the much-vaunted TV broadcasts, a piece of outreach in which NASA had a lot invested in terms of getting the American public firmly behind this extremely expensive tax-funded project. These reports from space were meant to last ten minutes every day, but Schirra unilaterally cancelled the first one and was clearly not going to let any 'dog-and-pony show' dictate the order of his working day. 'You've added two burns to this

---

* Actually, any illness in zero g is pretty grim, and I speak as someone who on the ISS had not a head cold but a urine infection with accompanying fever, the consequence of surrendering samples for medical analysis. I was more than usually grateful when the antibiotics eventually knocked it on the head. Incidentally, in retirement, Wally Schirra appeared in an advertising campaign for Actifed decongestant, the brand he had used in space. It made a change, I guess, from advertising Omega watches, which was the more typical post-flight early astronaut endorsement opportunity.

flight schedule,' he said over the airwaves, 'and you've added a urine water dump, and we have a new vehicle up here, and I can tell you at this point TV will be delayed without any further discussion until after the rendezvous.'

Without any further discussion? Were the limits of discussion actually Schirra's to decide? Was it not the job of the Flight Director to define where and when discussion ultimately ended? Schirra seemed to be boldly challenging the established order of doing things.

Flight Director Gene Kranz would later write, 'In the final days of the mission, the control teams, Capcoms and flight directors, covering for Wally, felt like embarrassed parents of a kid throwing a tantrum. Some of the exchanges seem sophomoric, but the stakes were high and discipline and team-work were victims of the feuding.'

A further line appeared to have been crossed when Schirra eventually told Mission Control, 'We're not going to accept any new games or do some crazy test we've never heard of before.' This was taking astronaut power to a whole new level – to the brink, indeed, of refusing orders, which in the military would have been unthinkable. And this was an ex-military crew, Schirra having been a naval pilot and Eisele and Cunningham having come to NASA from the Air Force. But, of course, spaceflight was not a military operation. It was a civilian operation. One that just happened to involve an awful lot of military people . . .

Anyway, that was only a *threat* not to comply. Actual non-compliance came in the final phases of the flight when Deke Slayton, in his capacity as Director of Flight Crew Operations, reminded the crew that, as a safety measure and as part of established protocol, they should put on their helmets for re-entry and landing.

The crew didn't want to do that. They said their head colds

made helmet-wearing uncomfortable and that they needed to be able to pinch their noses in order to equalise the pressure in their blocked-up ears, or else risk burst eardrums.

'If we had open visors, I might go along with that,' Schirra told Jack Swigert, the hapless Capcom charged with being the intermediary between the crew and the ground staff.

'OK,' Swigert replied. 'I guess you'd better be prepared to discuss in some detail when we land why we haven't got them on. I think you're too late now to do much about it.'

'That's affirmative,' said Schirra. 'I don't think anybody down there has worn the helmets as much as we have . . . We tried them on this morning.'

'Understand that,' said Swigert. 'The only thing we're concerned about is the landing . . . But it's your neck, and I hope you don't break it.'

'Thanks, babe,' said Schirra, dryly.

There was nothing Slayton could do. Schirra and his colleagues returned to Earth helmetless.

The typical Apollo-era hoop-la greeted their safe landing. Within days the three astronauts were being received at President Johnson's Texas ranch and appearing on *The Bob Hope Show*, in a special edition filmed in the auditorium of the Manned Spacecraft Center.

Nevertheless, behind the scenes post-mortems were being held and brows were being furrowed. Grumbling about 'Mickey Mouse jobs' was one thing, but refusing an order from Deke Slayton . . . well, 'mutiny' was clearly too strong a word for it, but this was certainly as close to mutiny as astronauts had ever strayed.

They did not go unpunished for it. Slayton, as we saw, held sway over crew selection. Schirra had already removed himself from the running by electing to retire, but Cunningham and

Eisele remained in the corps, and neither was picked to fly again. Only forty years later, in 2008, did NASA choose to award the Apollo 7 crew, in common with the crews of all the other Apollo missions, the Distinguished Service Medal. By that point only Cunningham was still alive. (He died in January 2023, aged ninety.)

Maybe all astronauts have ended up owing something to that Apollo 7 crew, though. With hindsight, their tetchy insubordination and general scratchiness over those eleven days in 1968 can at least be credited with significantly advancing debates started by the Apollo 1 disaster – debates around crew management, deferment, duties of care and what space agencies could reasonably ask of their astronauts in terms of risk and workload. In a sense, it obliged NASA themselves to confront another version of the 'astronaut as superhero' myth, so prevalent in the early media coverage.

A tacit assumption seemed to take hold that, because astronauts had been trained to travel beyond the limits that govern normal human life, then somehow they might be expected to operate beyond the limits that govern normal human behaviour too. The grumbling and grousing of the Apollo 7 crew might well have been 'sophomoric', as Kranz suggested, but at least it graphically revealed the unsoundness of that assumption and raised some important considerations for future flights. Humans everywhere in the world got sick and became tetchy. How reasonable was it to expect astronauts to be any different, just because they had stepped off the world for a while?

NASA suddenly had a few things to reflect upon in that department. And those reflections would only have to deepen as the missions they were sending their astronauts on grew longer and more demanding.

V. FISHING RODS AND EXTREME GARDENING

The astronauts referred to Skylab as 'the cluster', or sometimes 'the can'. Constructed in the wake of the Apollo landings, and opened for work in May 1973, the first American space station was built from bits of surplus Apollo hardware, lending it from the outset a slightly second-hand, patched-together feel.

The main workshop section was a reconstituted Saturn rocket booster, 48 feet long and just over 21 feet in diameter. It was divided into two decks, a large upper deck housing the laboratory and a low-ceilinged lower deck containing the crew quarters and kitchen. The lab also boasted a lavatory – or, as NASA formally referred to it, a 'waste management compartment'.* With the crew module docked onto one end of it, Skylab was 120 feet long, easily the biggest human-made object ever assembled in space.

Here was a massive step-change for the astronaut workplace. In a little over a decade, crew members went from flashing beyond the atmosphere and back in something the size of a hatchback to living up there for weeks at a time in a construction on the scale of a small warehouse. And, of course, the nature of the job altered and expanded accordingly, and with matching rapidity.

The Mercury and Gemini eras called for gutsy pilots, and Moon exploration needed gutsy pilots who were also scientists. The space station era would mean that astronaut work now called for gutsy pilots who were also scientists and who didn't mind doing a

* On the Space Shuttle, the 'waste management compartment' became the 'waste collection system' before morphing again on the ISS to become the still more effete 'waste and hygiene compartment'. According to my research, none of these terms is listed among the forty-eight *Oxford English Dictionary* synonyms for toilet.

bit of construction work and some maintenance engineering on the side.

Oh, and if they could also be multi-lingual and capable of some light dentistry and general practice doctoring, that would also be an advantage.

By the time the Space Shuttle was flying, crews were humorously hanging up signs referring to themselves as 'Ace Moving Company' (the crew of STS-5) and 'Ace Satellite Repair Company' (STS-41-C, which in 1984 flew into space to capture and repair the broken Solar Max satellite). Always open to broad definitions, 'astronaut' had evolved into perhaps the ultimate portfolio role.

But this brave new multi-tasking era did not dawn smoothly. Skylab was launched from Cape Canaveral and propelled into space on top of a Saturn V, the last use of that famous rocket. It was going to be inserted into orbit, around 265 miles above the Earth, and its first crew would fly up to join it and clock on for work twenty-four hours after it got there.

One minute into the rocket's flight the meteoroid shield tore off. This was not necessarily a ruinous development in itself because the chances of the station suffering a significant meteorite strike in its lifetime were vanishingly small. But the shield also doubled as a thermal control, protecting Skylab from the Sun's rays. With its heat-absorbent black and white covering gone, the gold foil on the outside of the station would be exposed and would simply soak up the heat. NASA calculated that, while the station was in direct sunlight, the temperature inside would rise as high as 77°C. The laboratory would quickly become uninhabitable, and therefore practically useless. And given that Skylab had cost $2.2 billion and taken about seven years to develop, nobody was terribly keen on that happening.

The loss of the shield also seemed to have caused damage to

the two main solar panels which were going to be providing the station with most of its power. When Skylab was finally in orbit, one of those panels only partly deployed and the other, seemingly trapped beneath bits of debris from the shield, didn't deploy at all.

Thus NASA found themselves effectively snookered: they had one problem requiring them to tilt the station into the Sun as much as possible to generate the maximum power from those limited solar arrays, and another problem needing them to keep a large part of the station in the shade to prevent it baking inside. Damned if you do, damned if you don't.

In any other circumstance, Skylab would have been hauled back into the garage for repairs. But that wasn't an option now it was in orbit. A team of astronauts was going to have to go up there to see what they could do about fixing it.

The three astronauts scheduled to be Skylab's first working crew were Pete Conrad, Paul Weitz and Joseph Kerwin. Conrad, just short of his forty-third birthday, was the veteran of two Gemini missions and had been the Commander of Apollo 12, during which flight he became the third man to walk on the Moon.* With a genial disposition and an engaging gap-toothed smile, his personal and often-repeated motto in the face of complex challenges was 'If you can't be good, be colourful'. Apollo 12 was frequently described as the happiest of all the Apollo missions – despite its rocket having been struck by lightning twice shortly after launch, creating a teetering pile of electrical glitches for the crew to cope with – and Conrad's seemingly irreducible cheerfulness was clearly at the heart of that. Even before

---

* Jumping from the Lunar Landing Module's ladder, Conrad, who was only just over 5ft 6in tall, shouted, 'Whoopee! Man, that may have been a small one for Neil, but that's a long one for me.'

the mission became additionally complicated, he was a sound choice to lead this fresh and uncertain experiment in long-term space-dwelling.*

Weitz, meanwhile, was a forty-year-old US Navy officer, piloting his first space mission; he would go on to command the Space Shuttle. And Kerwin, forty-one, was a former US Navy flight surgeon, appointed to this flight in the role of Science Pilot, NASA believing it would be valuable to have a medical doctor present on a mission which had among its prime objectives a study of the effects on the human body (and mind) of lengthy stays in space. The crew's stint on the station was scheduled to run for twenty-eight days – more than twice as long as any American had spent in space to that point.†

However, in the wake of the missing shield, the crewed launch was postponed and they were obliged to stay on the ground for the next fortnight while everyone got their heads together to decide how they could save the laboratory from an early trip to the cosmic scrapyard.

There were all sorts of alarming possibilities to consider. Could exposure to heat already have compromised the integrity of the station's structure? Could it even come apart before too long? Plenty for the crew to think about there, when they eventually stepped aboard.

---

* When Conrad died in 1999, following a motorcycle accident at the age of sixty-nine, NASA, as tradition dictated, planted a tree in his honour in the astronaut grove, just inside the gate of the Johnson Space Center. Each Christmas, the trunks of those trees are wrapped in white lights, except for one – Conrad's tree, which is wrapped in red, in an allusion to that 'If you can't be good, be colourful' motto.

† Three Soviet cosmonauts – Georgy Dobrovolsky, Vladislav Volkov and Viktor Patsayev – had already spent just over twenty-three days on board the Salyut 1 space station in 1971, but, as we shall see in Chapter Seven, their capsule depressurised on re-entry and they never made it home.

And assuming it remained in one useable piece, was the air inside it now poisonous? When it was in the full glare of the raw sunlight, the outside wall of the station had been exposed to searing temperatures, the sort of heat that could possibly have caused toxic vapours to seep from the station's insulation material. The crew now received additional training in the use of gas masks, which they were going to have to wear on first entry, while they checked the artificial atmosphere for toxins.

Beyond that, had the heat ruined the lab's long-term food supplies? Replacement food could be sent up in due course, but in the meantime the crew were given a crash course in food inspection, just to ensure they would know a dodgy mouthful when they saw, smelled or tasted one.

And what about the medical supplies? There were sixty-two different medications on Skylab and it was feared that half of them would now have spoiled. And film for the cameras? Had that already been ruined?

In the meantime, an emergency sunshade would have to be developed, something that could be erected over the exposed area of the station and do the thermal control job that the meteorite shield was supposed to do. Not for the first time, NASA's specialist departments now went into a kind of R&D overdrive. Special paint? Special wallpaper? Both were tested and ruled out. Some kind of weather balloon? Also tested, and ruled out because of fears that it would deflect heat to the solder in the joints on the solar panels and melt them.

At least three different improvised parasol prototypes were put forward, and after extensive trialling it was decided that the most practical of them seemed to be the one involving the fishing rods. Four fishing rods, to be exact – extendable ones, attached to a large piece of Mylar and nylon, 22 feet by 24 feet when unfolded. Once aboard the Skylab, the crew could poke the fishing rods out through the scientific airlock, a small hatch intended to be used for passing science experiments in and out of the lab. The rods could then be extended which in turn would unfurl the nylon so that it became a kind of protective wrapper against the Skylab's external wall in the damaged area.

Imagine putting up a tent, but through a letter box, and you will have some idea of the level of complexity, and the general air of the surreal, that was going to be involved in this piece of repair work.

Still, once packed down by professional parachute riggers, this cunning device could be fitted neatly into a 5-foot-long, 8-inch-square container – something greatly in its favour given the limited room for cargo in the crew's capsule.

Carrying the fishing-rod parasol with them, and additional food and medical supplies – and seemingly in very positive spirits, all things considered – Conrad, Weitz and Kerwin launched a fortnight later, in May 1973. After a seven-hour flight, the stricken space station came into view.

'Tally ho the Skylab,' Conrad cheerfully announced to Mission Control.

Conrad flew the capsule around the station to assess the damage visually and reported back. Things looked as bad as everyone had feared.

'Solar wing two is gone completely off the bird,' said Conrad. 'Solar wing one is only partially deployed.'

The gold foil on the outside of the lab module had already been blackened by eleven days of exposure to the Sun. In parts the surface looked bubbled.

To clear the debris that was stopping the solar panel from fully deploying, the crew had packed two tools – a long-handled tree-cutter and a multi-use tool with prongs and hooks, both acquired from a hardware store in Huntsville, Missouri.

One likes to imagine the conversation between the NASA

representative sent out to shop for these items and the store's assistant.

'Do you have anything for removing meteoroid shield debris from a space station's solar panel?'

'Hmm. Not really. Try Argos?'

Yet, as it happened, the hardware store *did* have something. Those tools were taken back to NASA's Marshall Spaceflight Center and lightly modified for use off the planet. Paul Weitz was about to do arguably the most bizarre and certainly the furthest-flung piece of gardening in history.*

The plan was to use the Apollo capsule as a kind of pick-up truck. Everyone put on their pressure suits. Once the capsule was depressurised, the hatch on the side was opened. While Conrad flew the capsule in as close as he dared, Weitz stood up in the opened hatch and tilted at the debris with the long-handled tools. Kerwin, meanwhile, stood inside the capsule holding on to Weitz's legs to stop him tumbling out and floating away into space.

It was dangerous work. It was hard work. But most of all it was frustrating work. Weitz managed to knock nearly all of the debris aside. But a metal strip with some bolts attached to it was proving impossible to shift. It was only a centimetre wide but it was wrapped tight around the solar panel and could not be dislodged. It was also too thick to cut. As Weitz precariously jabbed and knocked and tore at it with his hardware-store tree-cutter and his hooking tool, the communications feed with Houston became a blizzard of four-letter words.

Conrad had brought the capsule to within a couple of feet of

---

* On the subject of shopping around, you may care to know that the vacuum cleaner on the ISS is an off-the-shelf AC hand-held model from Walmart. I also retain fond memories, from just prior to my spacewalk, of fabricating a bolt-cleaning tool by fixing a spare toothbrush to a socket. It's not always rocket science.

the Skylab. At one stage Weitz's exertions jogged the Skylab so hard that its automated thrusters fired to correct its orientation. Fearing a collision, Conrad fired the CSM's own thrusters to pull it away slightly. At the time, Weitz's hooking tool was firmly clamped to something on the Skylab's panel, and the movement practically hoisted him out of the spacecraft altogether. Kerwin grimly hung on to Weitz's legs, narrowly sparing him the indignity (and extreme peril) of ending up hanging from a space station by the handle of a gardening implement.

Still the solar panel wouldn't come free.

Eventually Conrad had no option but to radio down to Houston with a report which will feel familiar to anyone who has ever finally stood back from an over-ambitious weekend DIY project.

'We ain't going to do it with the tools we got,' he said.

Weitz climbed down, and he and Kerwin closed the hatch and rejoined Conrad. They repressurised the capsule and exhaustedly removed their pressure suits. All they could do now was dock the spacecraft for the night, rest up inside it and then think again. What with everything – the early rise, the stress of the launch, the seven-hour flight, the suiting-up and the unsuiting, and, of course, the extreme gardening – it had been a long and debilitating day. All three of them were looking forward to recuperating.

But it wasn't over. Conrad guided the Command Module up to the lab and found that the capture latches on the docking adapter wouldn't operate.

He tried again. No joy.

He tried a third time. Then a fourth, a fifth, a sixth, running through all the variations of method available to him.

Still the capsule would not dock on to the Skylab.

There was one last option, short of giving up altogether and returning to Earth. They could get back into their pressurised suits, depressurise the spacecraft again, open the forward tunnel hatch, remove the back plate on the docking probe and bypass some of its circuits. Then they could use the thrusters to bring the craft into the docking adapter and hope that it worked.

This was a last resort, but it was at least one they had practised in Houston. That said, as they trained that day, Conrad had dryly remarked to Kerwin that if they ever reached the stage where this action was actually necessary, the mission was already doomed.

And now here they were, at that stage.

The three astronauts wrestled their weary limbs back into their spacesuits – itself a physically tiring process – depressurised, opened the hatch and set to work on the wiring of the docking probe.

Conrad then piloted the craft to the dock. This time all twelve latches engaged. Finally they could rest.

It had been twenty-two hours since they launched. Conrad, Kerwin and Weitz had just done the hardest day's work in space that any crew had ever completed to that point – and possibly since.

The following day, wearing gas masks, they entered the lab. It was extremely hot in places – too hot to work for long without taking breaks and heading back into the cool of the capsule. But, having established that the air was not toxic, they took off their masks and used the scientific airlock to deploy their twin-fishing-rod sunshade.

The crew were unhappy with the way it looked – it was wrinkled and not satisfyingly taut and smooth as they had hoped. But the team on the ground were whooping with delight. The Sun would heat the nylon and smooth out the wrinkles soon enough.

For Skylab, the healing could now begin. Two weeks later, Conrad and Kerwin were able to go outside the lab and finally cut through that last stubborn piece of debris, releasing the solar array. In their twenty-eight days on board, they did further repairs and found time to log 392 hours of science. Then America's first ever space station crew climbed back into their capsule and returned to Earth in triumph.

A month later, Alan Bean, the Apollo veteran, and two rookies, Owen Garriott and Jack Lousma, flew out to replace them. They took with them, among other things, two live minnows, fifty minnow fish eggs, six pocket mice, 720 fruit fly pupae and a pair of spiders named Arabella and Anita. Arabella and Anita were part of a student experiment to discover what happened to spider web-weaving skills in microgravity.*

In something of a related development, the crew also helped NASA resolve a question about humans and 'dominant verticals'. The docking adapter room on Skylab – the entrance hall, if you like, that astronauts first entered from their docked capsules – was a tube about 17 feet long, and deliberately arranged with no dominant vertical. Portals and equipment jutted out of its cylindrical walls at various points, in various attitudes.

This in itself was an experiment on NASA's part. If astronauts could happily operate in an interior with no ostensible floor or ceiling, simply floating to equipment as they needed it, with no regard for 'up' or 'down', then this was great news for future space station designers. It would mean that gear could be attached not just to the floors and walls of modules, but also to the ceilings, resulting in massive space-savings.

As it happened, the astronauts didn't seem to like the docking adapter room at all. Alan Bean confirmed a habit of orienting himself in a room according to its perceived dominant vertical. When he entered the workshop, for example, or the wardroom,

---

* After some initial confusion, leading to some very messy web-making, Arabella and Anita revealed that spiders can indeed weave their webs in space. Like astronauts, they just need some time to adapt.

where the crew prepared their meals, he preferred to think of those spaces as having a discernible floor and a discernible ceiling, even though, by the logic of space, there was no such thing. As Bean reported back to NASA, 'Now, if you want to put everything on the ceiling instead of the floor, we can sure handle that. It's just that we don't want half and half.' Orienting themselves within a room as they did on Earth was how astronauts seemed to feel most comfortable, so that's the way it stayed.*

What everyone did like was the window in the wardroom, looking out over the turning Earth. That window offered without question the best view of our planet that anybody in orbit had thus far seen.

If the first Skylab crew was remarkable for its positivity and its willingness to get stuck in on complex tasks in unpromising situations – and even using adapted gardening tools if necessary – the second crew excelled in its enthusiasm for the work they were set, willingly taking on extra jobs and even sometimes working through break periods. The joke around NASA afterwards was that Bean & Co. had accomplished 150 per cent of their mission goals. They also stayed in orbit for just over fifty-nine days. From NASA's point of view, this whole new 'living in space' thing was beginning to look easy.

But then came the third crew.

A little surprisingly, all three of the next team to inhabit Skylab were rookies. Commanders of the previous two Skylab missions had both walked on the Moon; they were pretty much the safest pairs of hands that NASA could muster. The third mission's Commander, Gerald Carr, was an ex-Marine who had been an astronaut for seven years, but he was unflown. So were his

* All the habitable spaces on the ISS are laid out to have 'floors' and 'ceilings' even though space laughs in the face of such a frail human concept.

crew-mates William Pogue, a former Air Force officer, and Edward Gibson, a physicist and a civilian astronaut.

They got off to the worst possible start when all three of them suffered space-sickness on arrival at the station. The first Skylab crew had all escaped nausea during their acclimatising period, and the second crew only had short and relatively light bouts of it. However, all three members of the third crew found themselves quite badly hit and throwing up.

At one point during this phase, Pogue vomited and Carr told him not to mention it to Mission Control, just to bag it up and throw it out of the trash airlock. This was in direct defiance of the regulations which stated that not only should all bouts of sickness be reported but that the vomit itself should be sent back to the ground for analysis.

They would keep it hushed up, said Carr.

'It's just between you, me and the couch,' said Pogue.

What they hadn't realised was that their conversation was on air, so that, as it happened, this exchange was just between Carr, Pogue and the whole of Mission Control. The following day the crew received a public reprimand from the ground – the first time in NASA's history that a formal ticking-off had been issued. It seemed to set the tone for relations between the crew and the ground from then on.

Even when they were over their initial sickness, the three astronauts seemed irritable, lethargic and prone to complain. Flight controllers had not seen this before from a crew. Gibson, for example, complained about the colour of the clothes they had to wear. Pretty much all of it was golden brown, the shade of the fire-resistant material from which it was made. 'I just get tired of this darn brown,' he moaned at one point. 'I feel like I've been drafted into the Army with this darn brown. It gets pretty obnoxious after a while. I'd like to get some different colour T-shirts.'

He also complained about the pockets. In an environment in which anything you put down floated away, space station clothes, necessarily, came with a lot of pockets. Gibson wished he had something 'without those blooming pockets, for comfortable casual wear'.

Carr and Pogue, being ex-military, were used to uniforms. But they found things to grouse about too. They complained that the 'waste management compartment' didn't have adequate footholds. 'You just ricochet off the wall like a BB in a tin can,' one of them said.* Carr, too, complained about the pockets – not from a fashion or comfort point of view, but because he felt they didn't fit the things they were intended for. The scissors pocket wasn't big enough for the scissors. The flashlight didn't fit in the flashlight pocket. You ended up putting things in pockets they weren't meant for. Carr grumbled to Ground Control that every time he bent to tie his shoelace he was in danger of jabbing himself in the groin with his scissors. He also complained about the mirror, in which he couldn't see his face adequately. 'There are better metal mirrors available than what we've got' was his verdict.

Also, the spoons they had to eat with were too small. 'We don't eat with tiny spoons on Earth – why do we do it here?' one of the crew asked Ground Control in exasperation. When asked to rate equipment on a sliding scale from good to poor, one of the crew ranked the spoons as 'lousy', adding, for the sake of complete clarity, that 'lousy' was 'somewhat below poor'.

As with the Apollo 7 crew, all three developed colds and nasal congestion. They sneezed a lot and needed copious quantities of nasal spray and tissues. They complained about the tissues, too. Gibson said that he wished proper handkerchiefs had been made

---

* A reference to the pellet from a BB gun, or air pistol.

available so that he wouldn't have to 'go around plucking tissues all the time'.

As the mission wore on, Carr and Pogue grew thick beards, and this too seems to have caused some consternation on the ground. Shaving in space had been possible since Apollo 10, using a razor and foam which held on to the hairs, preventing them floating off into the capsule, and which the astronaut could wipe away with a wet cloth.* The previous two Skylab crews had kept themselves relatively clean-shaven. But not Carr and Pogue.

NASA were bothered. Was all this moaning and seeming unhappiness the consequence of the atmosphere on Skylab? Were they now seeing some hitherto unrecorded mental effect of long-duration flight? Was isolation making this crew hostile? Was space having some kind of psychological effect on them that was making them behave this way? Or were these three crew members just . . . that way anyway? And were Carr's and Pogue's wild beards an indication that they had gone rogue? Or had they just . . . grown beards?

The crew's sickness meant their workload in the first week, which involved unloading and stowing the tonne of materials that had arrived at the station with them, was lightened to allow them to recover. From that point they were always slightly behind the curve, chasing to catch up, and at times they were clearly over-whelmed.

The assumption on the ground was that, after bedding in, the crew would simply pick up the mission where the previous crew had left it, and at the same pace. Consequently the third crew were being asked to do in their second week the kind of loads it

* The other great personal hygiene breakthrough on Apollo 10 was teeth-cleaning. From that point onward, a brush and paste were available. Until then a packet of Dentyne had stood in for formal brushing.

had taken the second crew – who were prodigious workers in any case – a month to build up to. Yet the more their performance deteriorated, the greater their workload seemed to become.

'On the ground,' Carr said plaintively one day, 'I don't think we would be expected to work a sixteen-hour day for eighty-four days, and so I really don't see why we should even try to do it up here . . . Please, loosen up.'

Of course, from NASA's point of view, almost the worst thing imaginable was the thought of three people in space with nothing to do. Every minute of this mission was costing thousands of dollars of public money; you couldn't have people just twiddling their thumbs up there, could you? The Flight Director, Neil Hutchinson, illustrated just how far apart ground and crew were on the matter of workloads when he later admitted: 'We prided ourselves here that, from the time the men got up to the time they went to bed, we had every minute programmed.'

In any case, hadn't the second crew lapped up the extra work and asked for more? Mission Control consequently assumed that filling the working day almost to bursting point would go down equally well with the third crew.

But it did not. According to Henry Cooper, reporting for the *New Yorker* in 1976, halfway through the mission Carr, Gibson and Pogue staged 'a sit-down strike' – although, of course, we should probably point out that weightlessness would have compromised their ability to do this, as indeed it would have prevented them from downing tools. But one gets the point. Gibson spent the day at the solar console in the docking adapter tunnel, looking at the Sun. Carr and Pogue went to the window and did some photography, taking pictures not of designated subjects but for their own amusement and pleasure.

Eventually, in a long and frank speech to the ground, Carr made clear his disappointment that he and his team were being made to work at the same pace as the second crew, something which he said he thought had been discussed and ruled out before launch. He also said he felt the crew were being kept in the dark about why their workload was rising. Was it because they had fallen behind? Or was it because NASA were forcing new projects into the schedule?

'That is essentially the big question, you guys,' Carr said. 'And that is, where do we stand?'

This time Ground Control seemed to respond to Carr's anguish. After this confrontation, the crew's workload lessened. Fewer experiments came on board, with more time allotted to complete them. In response, the astronauts' work-rate rose. In the later stages of the mission, something like harmony was restored and the mission was able to deliver on its targets.

As listed by Cooper in the *New Yorker*, the third Skylab crew supplied Earth resources data on 'ice movements, the migrations of fish, volcanic eruptions, melting snow, air pollution, water pollution, floods, droughts, sand dunes, sea conditions, weather conditions, hot areas in the ground which might provide geothermal power and floating seaweed that might be a source of food'. Its crew also set a record for long-duration stays in space that held for four years. As the capsule undocked for the journey back to Earth, even Gibson – formerly the complainer about the drab colour of the workwear and the lack of adequate handkerchiefs – sounded nostalgic about Skylab and about what he had just been through.

'It's been a good home,' he said over the radio.

Still, NASA had plenty of new things to think about. The fundamental question underpinning the entire Skylab programme was: could humans live in space for extended periods of time? But as the three missions progressed, it turned out to be a question with a perhaps unanticipated philosophical turn to it.

Because what was meant by 'live' in that sentence? Were NASA merely trying to find out whether people could *survive* in space? Or was there any sense in which living off the planet might be a fuller human experience than that? Was life in space only ever something to be endured? Or could it actually be enjoyed?

Interestingly, in the Skylab mission's development stages, a plan to allow alcohol on the station was briefly entertained: there was even a wine-tasting in Houston, so the astronauts could select their favourite for eventual stocking on board. Well, any excuse . . . But then anxieties appear to have grown about the message this would send and the plan was scrapped.

Pete Conrad, for one, thought that was a pity. If people on Earth could kick back and have a drink at the end of the day, he reasoned, why not people in space?

The point was, though, whether with a well-stocked cocktail cabinet or otherwise, it was clear that astronauts were now going to be spending longer and longer periods off the planet. And, given that this was the case, it was going to be necessary to start thinking quite differently about the kind of experience they were going to have up there. And it stood to reason that the astronauts themselves were going to want to have more to say about what happened to them during those longer stays.

In the meantime, an interesting outcome emerged when the health data from all three Skylab crews was available for comparison. Relative to their recorded physical condition at launch, the first crew was in the worst shape of the three on their return.

NASA had had stretchers standing by on the rescue ship for that crew after splashdown. But that was because, after an eighteen-day orbital mission in a Soyuz capsule in 1970, the three returning Soviet cosmonauts had appeared to need assistance getting out of the spacecraft. In fact, Conrad, Kerwin and Weitz all managed to walk without assistance across the deck of their rescue ship, albeit that Weitz was somewhat wobbly on his feet.

However, close inspection subsequently revealed that Conrad & Co.'s fitness levels had deteriorated far more after twenty-eight days than those of the second crew, who had been in space for fifty-nine days. The least affected, and the closest to their pre-launch conditions, were the members of the third crew, after just over eighty-four days. Was it possible, then – quite contrary to what NASA and probably everyone else who had given it any thought had suspected – that the longer you stayed in space the better you grew to tolerate it?

There would be greater opportunities to put that encouraging theory to the test in the ensuing years.

In the meantime, there was one resounding conclusion to be drawn and taken forward from the variety of those three Skylab crews – the can-do, have-a-go heroes of group one, the driven over-achievers of group two, and the stubborn and self-determining rebels of group three: there was clearly no such thing as a standard, interchangeable, one-size-fits-all astronaut crew, any more than there was such a thing as a standard, interchangeable, one-size-fits-all astronaut.

VI. COLLABORATIONS AND COLLISIONS

Helen Sharman was from Sheffield. She was a doctor of chemistry who, while studying, had also owned and maintained a 750cc Suzuki motorbike. In the late 1980s, when she was in her mid-twenties, she was working as a researcher for the Mars confectionery company, testing ice cream flavourings and emulsifiers, when she heard an advert on the radio: 'Astronaut wanted. No experience necessary.'

The ad had been placed by a venture called Project Juno, a collaboration between the Soviet government and a consortium of British businesses including British Aerospace and Interflora designed to stimulate trade and relations between Britain and the Soviet Union as the Cold War thawed into history. The plan was to find a suitable British candidate, fund them to travel to Moscow for cosmonaut training in Star City, and then launch them on a Soyuz rocket to the Mir space station for an eight-day science-based mission.

No Briton had been a flown astronaut before, and Sharman was intrigued – by the prospect of the training and the time in Russia as much as the idea of spending a week on Mir. She looked at the requirements and felt she could meet them all: aged between twenty-five and forty, science degree, fluency in two languages, good level of physical fitness, proven dexterity in manual work . . .

Her intrigue was shared by 13,000 other applicants, though, and Sharman did not rate her chances very highly.

Yet she was still on the list when it was filtered down to 150; and still on it when, after typically thorough physical and psychological testing, the 150 became four. Along with Gordon Brooks, a university lecturer, Clive Smith, a Royal Navy physician, and British Army officer Tim Mace, Sharman flew to Star City to begin an eighteen-month training regimen and a Russian language course.

At one point, the British funding fell apart, and with it the British component of the science that was meant to be done during the mission, and the project looked set to be scrapped. But the Soviet government stepped in to plug the gap and training continued.

Until the last, Sharman, reasoning that the agency would want the chosen astronaut and the back-up to be the same sex, assumed that the seat on the flight would eventually go to one of the men, but she was wrong. In May 1991, at the age of twenty-seven, she launched from Baikonur with two Soviet cosmonauts, Anatoly Artsebarsky and Sergei Krikalev, a butterfly brooch her father had given her and a photograph of the Queen.

Mir was still technically under construction when Sharman arrived. Sent into orbit in 1986 and eventually destined to be a six-module space station for science and Earth observation, it boasted an acoustic guitar for on-board entertainment but was not otherwise rich in luxuries. Thanks to a computer glitch (which Artsebarsky and Krikalev had been sent up to fix) its solar panels were misaligned, so the station was subject to frequent power cuts, a bit like Britain in the seventies. First the fans would shut down, causing the station to fall eerily silent, and then the lights would follow, leaving the place lit only by the dim glow of a solitary emergency fluorescent tube. Sharman and her four Soviet colleagues (Viktor Afanasyev and Musa Manarov were already on board when she joined) would just have to sit there in the gloaming – or, rather, float there – and wait for the power to come back on.

In another quirk, the daily wake-up call and the spacecraft's emergency alarm were programmed to make the same sound. Oversight or tactics? You decide. 'You'd wake up unsure if it was time to get up or if you were leaking oxygen,' Sharman said afterwards. 'It got us out of our sleeping bags pretty quick.'

During her week, she worked on the station's in-progress science projects, including research into crop growth using wheat seedlings and potato roots and analysis of protein crystals. And she took a call from Mikhail Gorbachev, the Russian President, who wanted to congratulate her in person on what she had achieved. By the time her capsule came to rest in Kazakhstan, Sharman had categorically – and, for some of us, very inspiringly – proved that astronauts could come from Britain, too.

Almost six years later, NASA astronaut Michael Foale's first impression of Mir – now eleven years old – was that it smelled of oil, a little like a repair garage, with a slight top-note of mould. His second impression was that it was cluttered, haphazardly jammed with equipment and snaking cables. To make your way weightlessly down its interconnecting tubes was, he said, like swimming headfirst down someone's oesophagus.

Foale arrived at the station in February 1997. He was born in Lincolnshire in England to a British father and an American mother. The latter had taken him as a child to see John Glenn's capsule on display in a museum in Minneapolis. Foale thought it looked like a dustbin. But near it hung a model of a futuristic space station, and something about it fired his interest.

Foale went on to study Natural Sciences at Queens' College, Cambridge, where he joined a dining society called the Cherubs. The society's initiation ceremony involved standing up before the other members and announcing what you intended to do later in life to glorify the Cherub name. A student called Stephen Fry vowed to wear the society's tie on television, and kept his promise almost immediately by representing the college on the TV quiz show *University Challenge*. According to Fry, Foale stood up and 'promised he'd be a real Cherub and fly to the heavens. We had no idea what he meant.'

What he meant was that he would apply to join the astronaut corps at NASA, train for five years, fly five Space Shuttle missions and, somewhere in the middle of all that, launch to Mir.

The idea of American and Soviet astronauts working together in space would have seemed fantastical in the heat of the race to the Moon. When Deke Slayton flew to Athens for an international space conference in 1965, he stood alongside Alexei Leonov and, through a translator, vowed to join him in orbit one day. But did either of them believe those words were more than rote diplomacy?

Yet, just a decade later, that meeting came to pass, the Apollo module of Slayton, Thomas Stafford and Vance Brand docking in space with the Soyuz of Leonov and Valeri Kubasov, and the two crews exchanging a symbolic handshake. In an era of détente, and with a growing desire on both sides to build hugely expensive space stations, the way forward was clearly collaboration and pooled resources. The Shuttle-Mir project sent seven NASA astronauts to Mir between 1994 and 1998, including Shannon Lucid from the 1978 draft, who spent 179 days on the station in 1996 with two non-English-speaking cosmonauts, and whose description of Mir as 'cosmic tumbleweed' will take some beating. Shuttle-Mir in turn paved the way for the multi-national project which was the International Space Station.

Still, in 1997, Foale was lucky there was still a Mir to fly to. Three months before he arrived the station had suffered an almost ruinous fire when one of the tanks used to burn cartridges of lithium perchlorate for the oxygen system went up in flames. This was during a hand-over period, when there were six crew members on Mir rather than the usual three, including another NASA astronaut, Jerry Linenger, and ESA astronaut Reinhold Ewald.

The station filled with smoke and flame and the fire cut off access to one of the two Soyuz modules docked on the station at that point, which were effectively the crew's lifeboats. 'We couldn't see at arm's length,' Ewald reported. 'I could not at that time have imagined that we go on with the mission.' Ewald remembers watching a jet of flame burning against the hull of the station and wondering just how much Mir's wall could take before it came apart.

How long the fire burned depends whether you turn for your figures to the Russian space agency at the time (ninety seconds) or to witnesses (fourteen minutes). There were reports of gas masks not working, and of fire extinguishers being too tightly strapped to the walls and irremovable. In some versions of the story, the fire was not actively extinguished by the crew, but simply burned itself out. Linenger, who was a doctor, had to treat some of his fellow crew for burns. Toxic smoke took hours to clear.

The residue from that fire was still on the walls when Foale arrived, delivered to the door by Space Shuttle *Atlantis*. But his time on the station would produce its own, equally threatening drama.

It was six weeks into Foale's mission. He had acclimatised to life on board, setting up his living quarters in the Spektr module, tying his sleeping bag to the wall there, and doing his best to get used to the constant noise of the motors rotating the solar panels while he tried to sleep. And he had commenced his science work: studying the circadian rhythms of sixty-four black-bodied beetles, and deploying a miniature greenhouse to examine the growth of rapeseed in microgravity. There was scope in his day for regular hours of exercise, and he felt healthy and clear-sighted, which he put down to getting a break from the paperwork he had to do in Houston.

On this particular day, Vasily Tsibliev, Mir's Commander, was using a TV monitor and two joysticks to fly a Progress cargo vehicle up to the space station for docking. It was exactly like playing a computer game except, in this case, instead of being only pixels on a screen, the Progress vehicle was a metal container weighing 7 tons and Mir was a thin-membraned artificial environment with three humans inside it operating in the void of space.

Foale was watching the approaching vehicle from a window. It seemed to him that the Progress was coming in at the wrong angle for the docking, and also travelling quite fast . . .

Suddenly Foale heard Tsibliev shout at him to get into the

Soyuz module – in other words, to head for the lifeboat. As he was scrambling through to the Soyuz, there was a massive thud and Foale felt the whole station shudder and twist. He would later relate that he thought in that moment that the station was about to break up, that he had looked at the thin aluminium walls around him and was 'just waiting for them to part' and expose him to the void beyond.

The emergency alarm was now sounding, and Foale noticed his ears popping. Above the shriek of the alarm there was a discernible whistling noise. The station was clearly depressurising, its precious oxygen seeping out into space.

The Progress vehicle had crashed through a solar panel and smacked into the outside of the Spektr module, punching a hole in it. In the Base Unit, Tsibliev checked the needle on the pressure meter. It was dropping towards 600 millibars. Below 540 millibars, the crew would begin struggling to stay conscious. He calculated that they had about twenty minutes in which to save the station and themselves.

Foale sat in the Soyuz and waited for Tsibliev and the crew's engineer, Sasha Lazutkin, to join him. He was fairly sure they would have to abandon the stricken station and attempt to return to Earth in the capsule. But he could hear them shouting to each other back in the laboratory. When, after a few minutes, his crewmates still hadn't come, he floated back up to the Spektr module to look for them.

He found them ripping out cables in a panicked attempt to isolate the punctured module from the rest of the station. Clearly they had no plans to leave as yet. Foale joined in to help. When they had finished, they went to seal off the passage to the leaking module. There was a thick panel for this purpose which had a valve set into it which might have proved useful later for equalising the pressure. Unfortunately it was intricately tied to the wall, and they lost precious minutes trying to free it. Giving up, they scrambled around until they found a thinner panel. When they put it over the gap, they felt it sucked into place by the pressure of the escaping air.

Foale's ears were no longer popping. They had now shut off the leak from the rest of the station and the immediate danger was over. But what now? By tearing all those cables out of the way, they had substantially unplugged the Spektr's solar panels, which fed electricity to the rest of the craft. As the station's stored power drained away, the fans stopped working, the lights went out, the central computer shut down and the radio went off. They were now in the dark, disconnected and adrift.

As the crew floated in the blackness it also became apparent that, as a result of the collision and the fact that no electricity was reaching the gyroscopes, Mir had begun to tumble. It was turning initially at the rate of roughly one full rotation every three minutes. Polar lights and stars were winding slowly past the windows. If they were going to have any chance of bringing the station back to life, they would first have to stop it spinning.

The crew now hatched a plan to make the station steady. Foale read the stars out of the window to get a sense of the craft's orientation as it turned, and he relayed that information to Tsibliev who periodically applied squirts of power from the thrusters in the Soyuz capsule in an effort to counter the station's motion. To their immense dismay, at first these efforts only seemed to make Mir turn faster. But they kept trying and soon the station was relatively stable again.

Now they would have to wait and hope that the Sun could do its work and kindle some power in the damaged solar arrays.

The crew sat tight in the dark and the silence. Finally, to their immense relief, the fans began to whirr again and the lights came on. The station was still alive. And so were they.

Life on Mir gradually resumed. The crew used the limited power supply to charge batteries. Foale's living quarters and all his belongings were now sealed off in the Spektr module. Tsibliev rummaged among the Soviet supplies and found him a tooth-brush. Now that radio contact was back up, Ground Control in Moscow could fire Mir's thrusters to keep it stable. The reduced power meant there was constantly condensation to be mopped up. They used underwear and other used clothing for the job.

Later, their mission at an end, Tsibliev and Lazutkin were replaced on board by Anatoly Solovyev and Pavel Vinogradov who were able to put on pressurised suits and perform a kind of internal spacewalk to restore the damaged cabling in the Spektr module. They found a thin layer of frost on the worktop counters but, after two months of exposure to the void of space, the module was otherwise in pretty good condition, and they were able to retrieve Foale's laptop computer.

Space Shuttle *Atlantis* returned to collect Foale in October. His four months on Mir qualify emphatically as one of spaceflight's tougher shifts – though not as tough as that of Tsibliev and Lazut-kin, who had endured both the collision *and* the fire that preceded it. The Russian Federation would keep Mir going for another four years, until March 2001, when, finally vacated, it was allowed to drop back into the Earth's atmosphere and burn up above the Pacific. By that time, construction of the ISS, begun in Novem-ber 1998 and based around a Russian-built central node that had a lot in common with Mir's, was well underway and the first crews had boarded. In 2003, Foale himself flew to the ISS to command Expedition 8. No collisions that time.

Built by five space agencies – American, Russian, Japanese, Canadian and European – a habitat the length of a football pitch was now in orbit 250 miles above the Earth, and a whole new era had opened for life in space and for the work of the astronaut.

VII. THE WINDOW ON THE WORLD

The first thing I ate in space was a bacon sandwich. And unlike with John Young and Gus Grissom on Gemini 3, nobody had needed to smuggle it on board.

'Welcome to space,' Scott Kelly, the ISS Commander, had said over the intercom while we three new arrivals were shutting down the Soyuz capsule after docking. 'What would you like for dinner?'

Kelly pulled the bacon sandwich from the 'bonus food' container that had arrived ahead of us, stocked with personal treats designed to make up about 10 per cent of our food intake during the mission. Moreover, my individual tuck (and here I can only imagine Young's and Grissom's hungry jaws dropping several inches) had been created for me by the pioneering chef Heston Blumenthal, the consequence of a kids' competition to design a menu for an astronaut. Kelly could equally well have dug out a serving of Operation Raleigh Salmon (inspired by my three weeks of kayaking across Prince William Sound in Alaska, where we supplemented our rapidly dwindling rations by trailing hooks behind the kayaks and catching fresh salmon), Sausage Sizzle (to evoke family holidays in Scotland, cooking by the river), Truffled Beef Stew (a posh dish Heston thought I might like for a special occasion) or Chicken Curry (who doesn't love a curry?). Oh, and also Lime Curd (I fell in love with Key lime pie during my NASA NEEMO mission to Florida) and Stewed Apples (my favourite). Delicious!

Actually, maybe a bit too delicious. After two months in space my sodium levels were through the roof. So I wonder how closely Heston actually adhered to ESA's strict low-sodium-diet regulations, which are put in place to try to reduce bone atrophy.

Still, I have to say, that bacon sandwich tasted good. OK, it was canned, and heated through in the galley's electric food warmer, but it was just what you wanted at the end of a long journey. Moreover, no Congressional inquiry followed this blatant act of in-space snacking, and there were no questions asked in the House of Commons. The only remotely scandalous thing generated by my consumption of that sandwich in a publicly funded environment was the drooling expression on the faces of Kelly and Mischa Kornienko, both of whom had been in space for months and who hadn't smelled the alluring aroma of cooked bacon for a while.*

But this was 2015 and, of course, space food, even without the involvement of Heston Blumenthal, had evolved as surely as spaceflight had. And a lot of that evolution was down to the work of Rita Rapp, someone to whom everyone who eats in space, bacon sandwich or otherwise, owes a hymn of thanks.

Rapp was a scientist in the aeromedical laboratory at Wright-Patterson Air Force Base and then joined the NASA Space Task Force at its inception. In the early years of the space programme she took blood tests from the Mercury astronauts, devised science experiments for the Mercury and Gemini flights and helped develop elastic exercisers for experimental use in space. She also, incidentally, witnessed every crewed launch of the Mercury, Gemini and Apollo eras, with the sole exception of Shepard's

---

* Yes, I let them both have a bite. Very reluctantly, though. Incidentally, although sandwiches no longer need to be smuggled, if you want to get a gorilla suit on board the ISS, you're going to have to go through unofficial channels. Scott Kelly managed it, with, it seems, a little help from his brother Mark, a former Shuttle Commander and currently Senator from Arizona. When Scott revealed the suit by starting his daily duties in it, not only did the House of Representatives' Appropriation Committee fail to take umbrage, nobody in Mission Control even noticed at first. When they eventually did, they found it as funny as the rest of us, fortunately.

maiden flight. And she only missed that one because a broken leg prevented her being there.

When the Apollo programme started, Rapp moved over to nutrition. Her game-changing notion was that food should be considered not as an add-on or an afterthought but as part of a space mission's hardware. She also saw how not just the food itself but the manner in which it was eaten was of enormous importance to the psychological well-being of the astronaut. Because the fact was, clever though they were in terms of space- and weight-saving, and as smartly as they met the messy challenges of weightlessness, the sachets of suckable gunk that constituted the first shots at space nutrition actually seemed to depress the astronauts – to the point, even, where they avoided eating rather than bother with them. The crews of Apollos 7, 8 and 9, for example, all recorded high levels of food refusal and came back significantly thinner than they had gone up.

Rapp's Apollo Food Systems team accordingly devised the Spoonbowl, implemented from the Apollo 10 mission onwards, which allowed the user to dip a spoon into flexible bowls containing rehydrated food, just as you might spoon up food from a bowl on Earth. Plus the packaging enabled the food to be in chunks rather than pastes or powders. The comforting associations of the simple, familiar, human act of spooning solid food from a bowl got a positive response from the astronauts, in a way that (surprise surprise) compressed cubes of toasted breadcrumbs held together by a starch and gelatin coating and designed to duplicate the effect of toast had not. It gave them back their appetite.

Rapp also developed thermostabilised foods that could be eaten from a can. And she became the liaison between the astronauts and the food menu. When Charlie Duke requested grits for the Apollo 16 crew, Rapp ensured that a form of grits flew with them. On Skylab, her sugar cookies were considered so covetable that they became a kind of currency between crew members, a bit like tobacco in prison.

This might all seem peripheral, but it really mattered. Malcolm Smith, NASA's Chief of Food and Nutrition, made an important point in an article for *Nutrition Today* in 1969: 'Progress toward extended extra-terrestrial exploration may be no faster than our progress with the problems of advanced food technology.'

By extension, then, it would be no exaggeration to suggest that Rapp's work on that food technology directly quickened the arrival of long-duration spaceflight. It certainly made the food tastier. When Rapp died in July 1989, they placed a plaque in her honour at the Johnson Space Center – with the unqualified approval, I would suggest, of every astronaut who ever flew.

The quality of space food bounded forward again with the Space Shuttle. STS-4 flew with a fridge-freezer on board, ostensibly for a science experiment. But it was travelling empty so the crew made sure it was loaded with broiled filet mignon and vanilla ice cream. All of that had to be consumed by the end of day two so the freezer could be turned over to science, so the crew spent the first hours of their mission fuller with beef and ice cream than any crew in history. When science had finished with the freezer, the crew fell on it again and repurposed it as a drinks chiller.

Eventually menus on the Space Shuttle would routinely include shrimp cocktail, Mexican scrambled eggs, beef stroganoff with noodles, ground beef with pickle sauce, potatoes au gratin, or chicken teriyaki, followed perhaps by butterscotch pudding or peach ambrosia. And to drink: peach-apricot drink, strawberry drink, tropical punch, tea with lemon, Kona coffee . . . Robert Crippen once reported enthusiastically that the irradiated steak on the Shuttle tasted 'like it had just come off the grill'. He was some way, clearly, from the tinned meat and potatoes which Helen Sharman gamely made her way through on Mir.

When I arrived at the ISS it was in its fifteenth operational year, and, though much about its environment remained rudimentary and faintly hospital-like, it too was a long way from the crowded and damp-smelling jumble of Mir. The ISS's first components were put into orbit in 1998. The first crew – Yuri Gidzenko, William Shepherd and Sergei Krikalev – boarded in 2000. Since then there have been some leaks, there have been damaged coolant pumps, and in 2007 a 2.5-foot rip appeared in one of the solar arrays which needed some urgent and highly risky repair work (by Scott Parazynski and Doug Wheelock). But the ISS has ensured that humans have been living off the planet continuously for – at the time of writing – twenty-three years and counting.

And it's not just a radical up-tick in the quality of the food that has made working in space on the ISS a more than simply endurable experience for the contemporary astronaut. It's the cumulative effect of the sixty years of experiment and learning that we've been encountering in these pages – experiment and learning about how space works, and how people work, and how people work best in space.

For instance, on the ISS, the daily work schedule, laid out on your iPad, would have a red line moving across it in real time. For all of the astronauts, that red line was the target of some very strong and opposing feelings. You loved it when you were riding with it, or even leaping ahead of it; you despised it when you fell behind it. It could deliver a striking picture of your day, and your workload, getting away from you. And it could possibly drive you mad.

So, a year before I flew, the 'pink box' system was introduced. On your timeline there were tasks in blue boxes which were priorities and which had to be completed at the set moment, in order to dovetail with other schedules elsewhere. And then there were tasks in grey boxes which you were expected to do on time but for which there was wriggle-room if for any reason you needed it, and about which you were welcome to have a conversation with the ground. But any task in a pink box was one you could do at a time of your choosing. It mitigated to some extent the power over you of the red line and restored to you some control over the ordering of your day, which people clearly need if they're to function efficiently and contentedly. Even astronauts. And perhaps especially them.

In the meantime, the heartening early inferences drawn by NASA from the first three Skylab crews – that the longer you stayed, the healthier you returned – have come quickly to look premature. Ahead of the hard push to Mars, we are still understanding the effects of long-duration spaceflight on human health in the longer term, and every astronaut in space is still themselves an experiment, as well as an experimenter, supplying valuable data for the study. What is definitively established is the central importance of compensatory exercise. At least two hours of working out per day – on the treadmill, the static bike and the piston-driven multi-gym – is locked into the schedule to compensate for the loss of muscle mass and bone density that naturally occurs in weightlessness. Consequently, it's possible, in at least some respects, to return from a long-duration mission more fit than you were when you left home. When Peggy Whitson got back from her nine-and-a-half-month spell on the ISS in 2017, she could bench-press more than her own body weight, something she had not been capable of before.*

---

* Whitson's 289 days in space was the longest single flight by a woman at the time. It was topped two years later by Christina Koch's 328 days on the ISS. As noted earlier, Koch will be a member of the 2024 Artemis II Moon mission.

All in all, working in space has evolved to a point that would have startled the first occupants of Mir and Skylab. On the ISS, the inevitable sensory deprivation – the enclosure, the absence of fresh air and weather, the removal from the world of human events, the dissociation that can be felt when night comes and goes sixteen times in every twenty-four-hour period – has the edge taken off it in ways the earliest space-station dwellers could only have dreamed of: it's not just the access to chocolate pudding cake, but the nightly phone calls to partners and friends, the weekly video calls with the family, the occasional movie screenings, the fortnightly check-ins with the flight psychologist . . . *

All this, plus no shopping, no laundry, and the easiest of commutes – from bed to work in practically no time at all. Since the pandemic we would know it, of course, as working from home – except that in this case home just happens to be a unique experimental microgravity environment 250 miles above everybody else's home.

As often as I could during my six months on the ISS, and frequently while cleaning my teeth, I would look out of the windows in the station's Cupola. That was Helen Sharman's advice to me before I flew: make sure you look out of the window. (It's not bad advice in general, actually.) And, as I watched, maybe the Gobi Desert would pass by, or the Himalayan mountain range, or the twinkling lights from the fleets of fishing boats in the Gulf of Thailand, or any one of the countless extraordinary vistas of our

---

* Those calls with the psychologist were always a little, shall we say, hurried in my case. 'Yes, all good, thanks. Moving swiftly on . . .' But I understood why they were there and one of my post-mission debrief recommendations was that astronauts should have a greater chance to get to know the flight psychologist before flying, to make those conversations easier when they are needed.

planet and the space that surrounds it which that unique window constantly gave on to.

And I would be aware that it's not given to many of us to see the world this way, and that the privilege of those views was a more than decent trade for the extreme dangers, the obvious deprivations and sometimes maddening complexities of working in space.

CHAPTER SIX

GETTING OUT

'Now it's time to leave the capsule if you dare.'

– David Bowie, 'Space Oddity'

I. THE POSTER

7 February 1984. Two hundred miles above the Earth, Bruce McCandless squeezes the jet-pack's hand-held gas throttles and propels himself out of the long payload bay in Space Shuttle *Challenger*'s roof and away into the vacuum. He adjusts his velocity until a gap begins to open up between himself and the Shuttle, which is at that point orbiting the Earth at 17,500mph.

He patiently moves out to 10 metres, and then 30 metres, and then, gathering confidence, on out to 50 metres . . .

Eventually he is 91 metres away from *Challenger*'s safe haven. Meanwhile, far below his feet, the Earth sweeps under him in its luminous blue shell. McCandless is completely untethered, the first human to cast off entirely from his spacecraft and float free in the void of space.

On board the Shuttle, the pilot, Robert 'Hoot' Gibson, points a Hasselblad camera out of the window, alters its angle to create a level horizon in the frame, just as he was taught in astronaut photography lessons, and clicks.

'As I was doing this,' Gibson will recall years later, 'I said to myself, "Wow, if I don't mess these pictures up, I'm going to get the cover of *Aviation Week*."'

And, yes, Gibson's picture will indeed make the cover of *Aviation Week* – will be rushed in less than a fortnight onto the front of its 20 February edition, in fact. But that won't be the only exposure this image gets. Gibson has just taken a photograph that will come to be known simply as 'the poster' and adorn the walls of countless classrooms, workshops and 1980s teenage bedrooms – and also, incidentally, the jacket of this book.

The image of that solitary human figure alone above the Earth continues to call out to us. Arguably not since Neil Armstrong's full-length shot of Buzz Aldrin standing on the Moon had one astronaut's photograph of another so breathtakingly encapsulated the spirit of exploration or so emphatically rendered in two dimensions humankind's instinctive zeal for pushing at our boundaries.

And not since the Apollo 17 astronauts drove the Lunar Rover 4 miles from the safety of their Lunar Module had any human been in such extraordinary isolation – so far from base, so far from home. For this, surely, was the ultimate spacewalk, the spacewalk that even spacewalkers would find themselves stopping to marvel at – a moment of almost overwhelming vulnerability involving a degree of exposure which even now, when you look at that picture, has the power to send a reflexive tingle through the palms of your hands and the soles of your feet.

Even before he flew untethered in the void, McCandless had lived through some remarkable moments. As a naval aviator, he was on board the aircraft carrier USS *Enterprise* when it was despatched to form part of the naval blockade erected to prevent Soviet nuclear weapons from reaching Cuba in the Missile Crisis of 1962. Four years later, at twenty-nine, he was the youngest member of the NASA draft of 1966, and three years after that he was Capcom in Houston when Armstrong and Aldrin walked on the Moon: that's McCandless's voice saying, 'OK, Neil, we can see you coming down the ladder now.'

Yet, despite that tantalising proximity to spaceflight in its golden period, his own career as an astronaut seemed to stall for a while. Half of his NASA class went to the Moon, but not McCandless, who would have to wait another decade and a half for his own chance to make history in space. In that time, he dedicated himself to a niche interest: ways in which astronauts could propel themselves outside their craft and in space. And that work would lead him ultimately to helping research and develop a 300lb back-worn nitrogen jet-pack, commissioned from Martin Marietta Aerospace.

NASA, with its well-established partiality to a truly thunderous label that could be slimmed down to make a deadly dull acronym, would call this device the Manned Maneuvering Unit, or MMU. And in 1984, when mission STS-41B came around, the Space Shuttle's tenth flight, McCandless got his chance to strap himself into it and see if it worked.

'You have a lot of envious people watching you,' said Mission Control as McCandless flew. 'Looks like you're having a lot of fun up there.'

In fact, he was freezing cold. His teeth chattered and he shivered. The cooling system on his pressurised spacesuit was really designed for astronauts at work, moving their muscles around and generating some heat. But, apart from operating the throttles, McCandless in his MMU was essentially still – basically twiddling his thumbs, albeit with his mind locked in steely concentration. And space is always either too hot or too cold to remain still in.

Also, it was surprisingly noisy. Out there in the vastness of space, McCandless thought he might know a moment of quiet communion with the void, but he was thwarted in that. As he put it: 'With three radio links saying, "How's your oxygen holding out?", "Stay away from the engines!" and "When's my turn?", it wasn't that peaceful.'

He didn't spend much time enjoying the view. He was concentrating on orienting himself in relation to *Challenger*, behaving like an aircraft flying in formation. But at the point at which he did allow himself the luxury of a glance down at the Earth, he caught a clear view of the peninsula of Florida and was able to feel in full how far he had come, in every sense.

After McCandless, the MMU was successfully flown on the same mission by his colleague, Bob Stewart, and on two subsequent Shuttle missions. In truth, though, its practical applications struggled to outweigh its manifest perils. In the space station age, spacewalks were maintenance missions, most safely achieved by rigorously planned tethered climbs along the structure of the station, or by capturing satellites, say, with a robotic arm and pulling them into the Space Shuttle's bay for an overhaul.

But the MMU, even its most ardent fans had reluctantly to concede, was a fun ride far more than it was an essential tool. It was fundamentally wedded to the spirit of getting out there in order to get out there – the jet-ski, if you like, of the spaceflight world. During the strategic rethink that followed the *Challenger* disaster, NASA decided not to allot budget to re-certify the MMU. It was scaled down into a smaller and much easier-to-wear unit – a kind of life jacket that an astronaut could turn to and give themselves a fighting chance if ever they became untethered and started to float away. NASA dubbed that one the Simplified Aid for EVA Rescue, or SAFER.*

But the MMU had yielded that photograph – a monument to

* Do you see what they did there?

226

outlast it. In the shot, the gold anti-glare visor on his helmet is down, and later McCandless would talk about how much he liked the picture's anonymity. 'It could be just about anybody in there,' he told *National Geographic* in the last interview he gave the magazine before his death in 2017. 'And I think that's part of the attraction.'

Indeed, maybe it cuts to the heart of it: that figure riding the void may well be 100 metres from the Space Shuttle, 200 miles above the Earth, flying as free as any human ever has, but the actual astronaut disappears inside the suit. It's an image of one of the most striking moments of individuality in the history of human aviation, but the human at the centre of it is not disclosed, allowing us to feel that it could be any one of us, or all of us.

And, when you think about it, the picture shares that generality with all the iconic images of astronauts boldly venturing outside their spacecraft and into the hostility of space, not least the famous shots of the Apollo crews on the Moon – those utterly unique human beings who, at the same time, stand before us in their suits and helmets as generic figures, showing us not themselves, but the reflections in their visors.

Still, *someone* has to get out of the spacecraft to bring us those pictures. And in that case it was Bruce McCandless.

II. THE GOLDEN EAGLE AND THE INFLATED ODDS

Of course, before astronauts could strap on a jet-pack and zoom through space, they had to learn to walk there. And given how the Soviet Union spent a large portion of the 1960s beating the United States to the big spaceflight landmarks, it will probably come as no surprise that the first person to achieve that complex and fraught manoeuvre was Russian. It probably won't be surprising, either, that the person in question almost died in the process.

As we've been discovering, among the many secrets of the success of the early Soviet space programme was a willingness to cut the occasional corner where it was felt expedient to do so. For instance, when Voskhod 1 flew, in October 1964, and made thirteen orbits of the Earth during just over twenty-four hours, it did so with a crew of three on board. To put this in perspective, America wouldn't feel it had the capacity to send three astronauts into space together in the same craft for another four years.

But the Soviets had done so by the simple space-saving expedient of squashing the crew into a capsule designed for two. In order to make room for one more cosmonaut, the decision was taken to leave behind pressurised spacesuits.

And if the cabin depressurised at any stage during those twenty-four hours? Well, that was just going to have to be a risk the cosmonauts must take.

However, the next Soviet mission, Voskhod 2, launched in March 1965, reverted to a crew of two, Pavel Belyayev and Alexei Leonov. And this time finding room for at least one pressurised spacesuit was non-negotiable: during this flight, Leonov was going to attempt to post another Soviet first by leaving the capsule.

It would not go smoothly. Indeed, it would turn into a fight for his life.

Before he became one of the true legends of Soviet spaceflight, Alexei Leonov seems to have had his heart set on becoming an artist. He was a student of drawing and painting at the Academy of Arts in Riga. However, on graduating he went to Air Force School in Kharkiv, Ukraine, where he qualified as a parachute instructor, and shortly after that he was recommended for the first cosmonaut draft.

As mentioned before, Leonov was slightly too tall to meet the

5ft 7in cut-off and join Yuri Gagarin in the Vanguard Six, but the Voskhod capsule would certainly fit him, as would, it was hoped, his Berkut spacesuit, although exactly how well wouldn't be known until he was actually in space inside it and it was fully pressurised.*

Still, in preparation for this world-first spacewalk mission he had undergone eighteen months of training, most of it snatched in the brief seconds of weightlessness available to him during parabolic flights on the 'Vomit Comet'. He was as ready as he would ever be.

Yet you may well ask: why? Why spacewalk? Why would you want to open the capsule of an orbiting vehicle and scramble out? Why would you take your chances in an atmosphere so hostile that in the absence of a spacesuit it would kill you on contact? Wasn't spaceflight, with all the doors firmly closed, dangerous enough already?

Well, clearly there were potentially practical reasons for mastering the art of spacewalking, even before the invention of space stations and orbital telescopes (such as the Hubble) that might need fixing. For instance, it would obviously be handy for astronauts to have the wherewithal to repair the outside of their spacecraft in an emergency – to know that, if the eventuality arose, they could get out and get under, as it were.

Spacewalking would also mean they could deploy useful equipment outside the craft, as and when necessary. Leonov's first listed task on exit from the Voskhod capsule was going to be to attach an external camera to it – not a critical job by any means, but plausibly a rehearsal for far more substantial acts of in-space DIY later on.

Yet, in the ragingly pioneering atmosphere of 1965, one can't

* 'Berkut' is Ukrainian for 'golden eagle'.

help but suspect that these sober, practical considerations were running second to that most compelling of arguments for getting out of a spacecraft in the ruthlessly hostile void of space – namely, because you could. Or, at any rate, because you *possibly* could. (Discussions on this could resume when, and if, Leonov returned.)

Perhaps more sharply than at any other moment in the development of spaceflight, we see curiosity firmly in the driving seat here. I mean, you wouldn't go all the way to the Grand Canyon and not get out of the car, would you? Similarly, you wouldn't go all the way into orbit and not take a little turn outside the spacecraft. That seems to have been the thinking. Spacewalking happened for no better reason than because, having gone all that way, it would have been a shame not to.

The rocket containing Leonov and Belyayev left Baikonur shortly after dawn on a Thursday morning in March 1965, and ninety minutes into the flight and in orbit 200 miles above the Earth, Leonov got ready to climb out.

There was no means of depressurising the Voskhod in flight, so in order to transition to the vacuum of space Leonov had to use an expandable rubber airlock, deployed like a kind of old-fashioned beach changing-hut around the outside of the hatch. Known as the Volga airlock, this device had been designed, tested and pressed into service in just nine months. Once inside this little cubicle, Leonov checked his suit, adjusted the air pressure, attached himself to a thick, 26-foot-long tether and thrust himself out into the vacuum.

'Nothing will ever compare to the exhilaration I felt,' Leonov said of the experience of becoming the first human to see the bright blue world and the universal blackness around it through only the thickness of his own visor.

But exhilaration would not be his only emotion during that pioneering twelve minutes and nine seconds of floating outside the capsule. Filled with oxygen, his spacesuit greatly expanded and stiffened, converting Leonov into an unjointed doll and making movement inside the suit's hard shell almost impossible. His hands slipped back within the engorged fingers of his gloves and his feet withdrew from inside his bloated boots, and his arms were so restricted by the padding around them that he couldn't reach round to the controls on the camera that was strapped to his chest.

Still, he was outside the capsule! He was spacewalking!

Leonov let the tether pay out and drifted away backwards into the void. Then he used the tether to haul himself back to the capsule again. He did this a couple of times, enjoying the sensation of it – a sensation he was the only human alive to have known.

It was when he tried to clamber back through the hatch into the airlock that Leonov's job suddenly got harder. He was now too swollen in the suit to fit back in. He squeezed and pressed himself downwards, working himself into a lather of sweat, but to no avail. He was now a cork attempting to re-enter a champagne bottle. And if he couldn't find a way to solve this predicament, he would be stuck outside the spacecraft until his oxygen ran out.

So, at vast risk to himself, Leonov reached for a valve on a part of his suit that he could still get a hand to, and opened it to partly depressurise the suit.

This move, he later claimed, had always been part of the plan, but it was still a wildly dangerous thing to do – an act which could have deprived Leonov of oxygen and could also have given him an instant and lethal dose of decompression sickness, or, as divers know it, 'the bends'.*

---

* In decompression sickness, or 'generalised barotrauma', nitrogen in your bloodstream is given insufficient time to clear and separates out of your blood

Still, having successfully made himself smaller, Leonov was then able to wrestle his now hot and aching body all the way back in and, after still more wrestling, close the hatch behind him.* In the course of these exertions, he lost 12lb of body weight and reported that, by the end of the ordeal, sweat was actually sloshing around in his boots. Reviewing this escapade from the other side of the world, Gene Cernan declared Leonov 'one of the gutsiest men alive' – and, as we shall see, Cernan himself would soon enough know quite a bit about spacewalks that didn't go exactly according to plan.

After all that, a smooth return to Earth for Leonov would have been nice. No such luck, though. Still to come: wolves and bears. First, though, the act of ejecting the no-longer-needed external rubber airlock seemed to send the Voskhod capsule into a sickening spin which it took Leonov and Belyayev several alarming minutes to correct. Then the capsule's primary retro-rocket misfired, necessitating an entirely manual re-entry involving a steep descent angle and ruthlessly punishing g forces, which eventually saw the capsule come down 600 miles off course, in the Ural mountains.

Once they came to rest, Leonov and Belyayev triggered the craft's hatch but discovered it would only open a little way. Inconveniently, the capsule was wedged tight against a birch tree. The cosmonauts rocked back and forth in the craft until they had

---

and forms bubbles. It's known colloquially as 'the bends' because it seizes up the joints and the pain is acute enough to double you over. It can also kill you.

* Some versions of this story have Leonov entering the airlock head-first, rather than feet-first as planned, and thereby having to perform a contorted flip inside to get the hatch closed, increasing his discomfort. But that is contradicted both by notes Leonov made after the flight and by filmed footage. This is not a story that really needed any whooping up, but it seems to have got some anyway.

twisted the hatch away from the obstructing tree-trunk, and finally it sprang open, admitting a blast of bitterly cold air.

Leonov and Belyayev jumped down and promptly sank up to their knees in snow. Leonov realised that the sweat he had shed during his spacewalk exertions was still soaked into his clothes and his boots. If he now allowed it to freeze he would be dicing with hypothermia. So, ignoring the biting cold, he stripped naked, emptied his boots of their liquid, wrung out the rest of his under-garments, and then put everything back on again. He and Belyayev then spent a frigid night camped in the capsule, listening to the daunting cries and roars of the hungry-sounding local wildlife, while the temperature outside dropped to an inhospitable minus 22°C.

The following morning a rescue party, including two doctors, reached the cosmonauts on skis. But all of them would have to spend another night on the mountain with the bears and wolves while a clearing was hacked out of a portion of forest nearby in which a helicopter could land. That second morning, the cosmonauts skied 9 miles to the freshly improvised heli-pad and were finally airlifted back to Baikonur.

Happily, one thing that survived all this turmoil was a drawing Leonov had made in the capsule. Somewhere between becoming the first person to leave their space capsule and piloting himself back through the atmosphere, Leonov had found time to do some sketching, producing an impression of an orbital sunrise. Technically that was the first work of art produced in space, although arguably Leonov's spacewalk was another and perhaps greater one.*

* The legacy of Voskhod 2 lives on. To this day all ISS crew members undergo two nights of survival training at temperatures well below freezing with just a parachute for shelter. We're generally spared the wolves and bears, though.

III. 'I'M NOT COMING IN!'

The original plan for what NASA were now referring to as an 'Extravehicular Activity', or EVA, on Gemini 4, set to launch in June 1965, was a simple one. Ed White would stand up and put his head out of the hatch, while James McDivitt, down below, held on to his legs.

Or, as McDivitt explained it to the mission-announcement press conference: 'Ed White will probably be the standee, and I probably will hold on to him.'

'Yes,' said White. 'I'm the standee, and he is the holdee.'

But, of course, Alexei Leonov's adventure that March – twelve minutes of full-body floating, captured on camera – made merely poking your head out of the door of a spacecraft look . . . well, a bit lame. Those pesky Soviets had done it again. NASA's plan for a full EVA was therefore rushed forward, and was ultimately put together with only a week of rehearsal time for the astronauts.

Still, White and McDivitt were a promising team for what was clearly going to be a slightly on-the-hoof job. They knew each other from the University of Michigan and had graduated together from the US Air Force Test Pilot School. McDivitt called White 'the best friend I ever had'. White, for his part, was a talented athlete, a 400m hurdles specialist who in 1952 had narrowly missed out on selection for the US Olympics team. Now he would attempt to go for gold in the newly opened spacewalk category.

Once he was in orbit during the mission, the checklist for the EVA's equipment preparation was fifty-four items long. White methodically worked his way through it. Then, over the Pacific Ocean somewhere above Hawaii, he pushed open the Gemini's hatch – after a tussle with its apparently sticky catch – and thrust himself out into the vacuum of space.

Alan Bean on the Moon during the Apollo 12 mission, November 1969. For all of the Apollo astronauts who had the experience of walking on the Moon, the pressure was now on to account for it. 'What was it like?' and 'How has the experience changed you?' the public asked. It was Alan Bean's plausible contention that 'everyone who went to the Moon came back more like they already were'.

The wives of Apollo 12 astronauts: Sue Bean, Barbara Gordon and Jane Conrad showing their support during the Apollo 12 mission, November 1969.

Mission Control Flight Directors – Gerald D. Griffin, Eugene F. Kranz and Glynn S. Lunney – celebrating the successful splashdown of the Apollo 13 crew. Gene Kranz thought returning the Apollo 13 crew alive was 'NASA's finest hour', eclipsing even the Moon landing. It would be hard to argue with him.

New beginnings, July 1975: US astronauts Thomas Stafford and Donald Slayton hold containers of Soviet space food in the Soyuz Orbital Module during the joint US-USSR Apollo-Soyuz Test Project – a symbolic handshake, where the two craft docked in orbit. Vodka labels have been pasted onto the containers for a toast to the other crew.

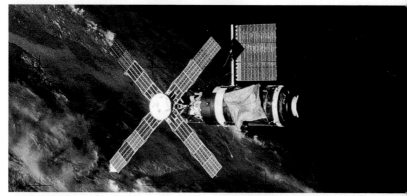

Astronauts referred to Skylab as 'the cluster', or sometimes 'the can'. Constructed in the wake of the Apollo landings, and opened for work in May 1973, the first American space station was built from bits of surplus Apollo hardware, lending it from the outset a slightly second-hand, patched-together feel.

Beards of protest? 'On the ground, I don't think we would be expected to work a sixteen-hour day for eighty-four days, and so I really don't see why we should even try to do it up here . . . Please, loosen up.' The complaint levelled at Mission Control by the third crew of astronauts to spend time aboard Skylab.

The Space Shuttle *Atlantis* docking with the Russian Mir space station in 1995. Mir was the first modular space station and was constructed in orbit between 1986 and 1996. The Shuttle-Mir programme would eventually include eleven Space Shuttle flights.

In 1997, the Mir space station suffered an almost ruinous fire when one of the tanks used to burn cartridges of lithium perchlorate for the oxygen system went up in flames. There were reports of gas masks not working, and of fire extinguishers being too tightly strapped to the walls and irremovable. NASA astronaut Jerry Linenger (pictured in his gas mask during the crisis) was a doctor, and had to treat some of his fellow crew for burns. Toxic smoke took hours to clear.

Helen Sharman returning to Earth after her 1991 journey to the Mir space station. Sharman had categorically – and, for some of us, very inspiringly – proved that astronauts could come from Britain, too.

The ground-breaking 1978 astronaut draft, the Thirty-Five New Guys, which properly opened the doors for women and civilians, was made under George Abbey's supervision. It put the first American woman in space (Sally Ride), the first African American in space (Guion Bluford) and the first Asian American in space (Ellison Onizuka), and it produced the first American woman spacewalker (Kathy Sullivan).

The first American woman in space. 21 June 1983, astronaut Sally Ride, STS-7 mission specialist, in the mid deck of the orbiting Space Shuttle *Challenger*.

The Space Shuttle era. Sally Ride and her other crewmates rocket into space aboard the Space Shuttle *Challenger*, June 1983.

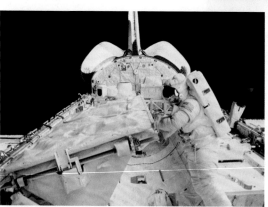

Kathy Sullivan becomes the first American woman to perform a spacewalk on the Space Shuttle.

The *Challenger* crew. The Space Shuttle project never recovered from the tragic loss of *Challenger*. Its scope was narrowed, its blithe certainty about itself gone and unrecoverable.

One thing that survived intact in the *Challenger* wreckage was a football. It had been packed for the journey by Ellison Onizuka, and was the ball used by his daughter's high school team. Thirty years later, the NASA astronaut Shane Kimbrough asked if the ball could accompany him on his mission to the ISS. He took a photograph of it floating against the backdrop of the Cupola observatory module window.

Landing of the first Space Shuttle mission, 14 April 1981. The rear wheels of the space shuttle orbiter *Columbia* touched down on Rogers Dry Lake at Edwards Air Force Base, NASA's Armstrong Flight Research Center.

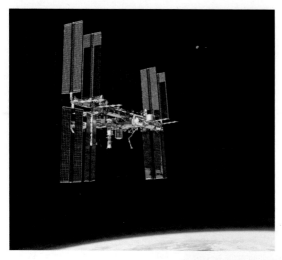

July 2011. A photograph of the International Space Station taken from aboard the Space Shuttle.

December 2006. With the islands of New Zealand as a backdrop, astronaut Robert Curbeam Jr (*left*) and European Space Agency astronaut Christer Fuglesang of Sweden participate in an STS-116 spacewalk.

My proudest moment: my EVA from the ISS on 15 January 2015. About a third of astronauts have been fortunate enough to perform a spacewalk or EVA (Extravehicular Activity).

I was just as proud to be part of a six-month mission aboard the ISS as part of an international team of astronauts. The human collaboration aboard the ISS may well be remembered as one of the station's greatest legacies.
Top to bottom on our launch day: Yuri Malenchenko (Roscosmos), Tim Kopra (NASA) and me (ESA).

April 2023. The Artemis II crew announced: Christina Koch, Victor Glover, Reid Wiseman and Jeremy Hansen. As NASA Administrator Bill Nelson says, 'The Artemis II crew represents thousands of people working tirelessly to bring us to the stars. Together we are ushering in a new era of exploration for a new generation of star sailors and dreamers – the Artemis generation.'

The Space Launch System rocket, the most powerful rocket in operation, will launch NASA's Artemis III crew on a journey to the surface of the Moon in the next few years. Lunar flight is back on.

The suits of the future. *Left to right*: NASA astronauts Jessica Watkins, Robert Hines, Kjell Lindgren and ESA astronaut Samantha Cristoforetti sitting aboard the SpaceX Crew Dragon, which flew and docked with the International Space Station in April 2022.

He was attached to the Gemini by 25 feet of gold-coated umbilical, feeding oxygen into his suit. He was also gripping a Hand-Held Maneuvering Unit, or HHMU, a pistol with three small rocket nozzles that could squirt compressed oxygen at the push of a rocker-switch and help him project himself through the void.*

Because the capsule had passed briefly out of communication range, Mission Control missed the start of the action.

'Has he egressed?' Gus Grissom, acting as Capcom, asked when the signal was re-established.

McDivitt answered, 'He's out, Gus, and it's really nifty.'

White, it was clear, was having a high old time. 'This is the greatest experience,' he shouted. 'It's just tremendous. Right now I'm standing on my head and I'm looking right down, and it looks like we're coming up on the coast of California . . . I'm not coming in! This is fun!'

He drifted out to the tether's full extent and then returned to the craft three times.

It was a struggle, though. The tether kept tugging him towards the back of the capsule, where he was at risk of getting squirted by the craft's automatic thrusters and doused in an unpleasant mix of monomethyl hydrazine and nitrogen tetroxide. Also his direct communication with Mission Control – operating in Houston for the first time rather than, as previously, from Cape Canaveral – wasn't functioning; he could only hear McDivitt in the capsule who had to pass on instructions from the ground.

Still, this was definitely, as McDivitt had said, nifty, although a stark reminder of the consequences of coming adrift was delivered when one of White's spare thermal over-gloves rose from the

---

* Sandra Bullock would attempt her own version of this trick many years later in the movie *Gravity* (2013), but using a fire extinguisher in place of the HHMU. Don't try this at home. Or, indeed, in space.

capsule, departed the hatch and drifted past him into space on a walk of its own.

'It looked like it was on a definite trajectory going somewhere,' White later said about watching the glove pass over his shoulder and escape for good. 'I don't know where.'

In an extraordinary oversight, nobody had thought to attach anything to the outside of the Gemini that White could grab at when he needed to, so his attempts to steady himself against the capsule every time he returned to it tended to result in ungainly, flailing limbs.

'There just isn't anything to hold on to,' he said in exasperation at one point.

Occasionally McDivitt would hear White's boots clump against the outside of the capsule, and at one point he watched as his colleague wiped himself slowly across the window.

'You smeared up my windshield, you dirty dog,' said McDivitt.

After fifteen minutes, McDivitt, in his enforced role as translator, asked Houston if there was anything they needed.

'Yes,' said Flight Director Chris Kraft. 'Tell him to get back in.'

White reluctantly obeyed.

'I'm coming back in, and it's the saddest moment of my life,' he said, mournfully.

At the hatch, he handed down the zip-gun and his camera to McDivitt and undid his connections. Then, with McDivitt pulling on his feet and dragging them into the footwells, and pausing occasionally to fight off attacks of cramp in his legs, White forced his way back down into the capsule. Another struggle then ensued with that reluctant hatch, McDivitt holding White down while White yanked at the catch, though not so hard as to risk breaking it, because, as they well knew, if they couldn't get the catch to work, they wouldn't be going home. This took a while and a lot of suited-up effort from both of them. Eventually, though, the catch responded to their coaxing and clicked into place. Soaked in sweat, the pair of them then spent an orbit in their seats recovering their breaths before repressurising the spacecraft.

White was exultant, but drained. It later emerged that the fuel in the HHMU had run out after only three minutes. All of his motions for the next twenty minutes had been accomplished by his own muscle-power, hauling on the tether and twisting himself inside the stiff suit – a major full-body work-out. Still, White had not only enjoyed himself more than Leonov, he had been outside the spacecraft for almost twice as long, delivering a much-needed boost to NASA's pride.

He had also taken some good photographs, although it appears to have required a little effort to convince some people at NASA of their merit. Part of White's brief was to take pictures of the outside of the craft which had suffered some minor damage during the launch phase. But he also shot pictures of the Earth beneath his feet, including Africa as it wound by below him. Though the engineers were pleased with the technical shots of the craft, the other images didn't seem to grab them. Certainly the initial verdict of NASA's Bob Gilruth was less than enchanted. 'Those are just pictures of the Earth,' he said.

Well, I guess it's like they say: when you've seen one Earth, you've seen them all.

Nevertheless, White's spacewalk marked the moment when photography's importance for spaceflight really began to come into its own. A man called Richard Underwood led the small advisory group responsible for photography on NASA missions, and trained up the Gemini astronauts in the use of cameras. Underwood would tell them, 'Your key to immortality is in the quality of the photographs, and nothing else' – and he may not have been entirely exaggerating. Those astronauts could bring back all the data in the world, but in terms of communicating with the world, their pictures would define them.

And spacewalking was already revealing itself to be the best photo opportunity of them all.*

IV. WRESTLING AN OCTOPUS

The next American to walk in space was going to be Charlie Bassett. Bassett was a thirty-four-year-old former Air Force Major, assigned to Gemini 9 in November 1965 alongside Elliot See, thirty-eight, a civilian test pilot. The pair were deep in training for an unprecedented three-day mission that was going to involve a docking procedure with an Agena target craft and, for Bassett, a spacewalk incorporating tests for a new propulsive backpack, the Astronaut Maneuvering Unit, designed to make getting around in the void that much easier than the zip-gun that Ed White had fleetingly been able to use.

At the end of February 1966, less than three months before their scheduled launch, Bassett and See, along with their back-up crew Gene Cernan and Tom Stafford, were due to pay a routine visit to Lambert Field, St Louis, where their spacecraft was being prepared in a warehouse belonging to the McDonnell Aircraft Corporation, prior to getting shipped down to the launch site. The four astronauts jumped into a pair of NASA's T-38 Talon jets for the ninety-minute hop to Missouri, Bassett and See taking the lead jet with the tail number NASA 901, Stafford and Cernan following in NASA 907.

It was a wet and foggy morning in St Louis, and the weather

---

* During my own EVA from the ISS, which we'll explore later in this chapter, I couldn't resist the urge to take a quick selfie, with the Union Flag on my suit taking its first walk in space.

was worsening, with thickening cloud and squalls of snow. When both the jets, flying in formation, with Stafford just behind See's right wing, dropped below the blanket of grey cloud, they were already low over Lambert Field's southwest runway and flying too fast to land. Stafford immediately ascended, ready to go around and approach again. See, for some reason, did not.

Instead, NASA 901 performed a tight right turn towards the runway. Cernan, sitting in the back seat behind Stafford, heard him say, 'Goddamnit, where's he going?'

See had no chance to execute a landing. Instead he found himself flying directly at the McDonnell building. He pulled the plane up and hit the afterburners but there wasn't time. The T-38 clipped the building's roof, shedding its starboard wing, and somersaulted into the car park beyond, where it exploded on impact, killing Bassett and See immediately.

Remarkably, nobody else died in this accident, although inside the impacted building debris dropped from the ceiling and a dozen workers received injuries. Had the jet hit the warehouse full on, rather than only clipped it, it could have been the end for numerous workers inside, for the Gemini spacecraft and possibly for the entire Gemini project.

Bassett and See were not the first NASA astronauts to die in a T-38 jet crash: Theodore 'Ted' Freeman from NASA's 1963 draft died in October 1964, ejecting too close to the ground after his T-38 suffered the misfortune of a birdstrike on its final approach to Ellington Air Force Base. Then, the year after Bassett and See's accident, in 1967, Clifton 'CC' Williams was killed when his jet suffered a mechanical failure on a trip home to see his parents in Mobile, Alabama. In 1972, two NASA test pilots, Stuart Present and Mark Heath, lost their lives in a T-38 attempting an approach in fog. Several other astronauts experienced near misses in those planes, including Ed White, Rusty Schweickart, Pete Conrad and

Ken Mattingly. The early astronauts loved their jet-flying but they paid heavy prices for it.

Bassett's Gemini mission fell to his back-up, Gene Cernan. As the second American to attempt to walk in space, Cernan had the immense reassurance of Ed White's prior experience to draw on. Or maybe not. 'It can be safely said that we didn't know diddly squat about walking in space when I popped my hatch open on Gemini 9,' Cernan later wrote.

Walking in space, of course, was almost impossible to practise without access to space itself. The Gemini astronauts trained on a shiny metal floor, about the size of a boxing ring, on which they rode around on something that Michael Collins compared to a floor-waxer, squirting their zip-guns to direct themselves. This sounds like good fun, and probably was, but as Collins also pointed out, such a device enabled you to practise only six of the twelve directions of movement available to you in space. So, not the best of simulators.

The Gemini astronauts also, like Leonov, used the zero-g aircraft but, as Collins further pointed out, that only gave you about twenty-eight seconds at a time, and with a lot to cram in, it mostly led to 'pandemonium . . . 28 seconds of slashing around'.

So, we can fairly confidently assert that it was a significantly under-rehearsed Gene Cernan who radioed down to Mission Control on the third and final day of the mission and said: 'This is Gemini 9. We are now going to walk.'

Pushing his head through the hatch, Cernan felt overwhelmed – just as Alexei Leonov had done – by what he saw. 'Nothing had prepared me for the immense sensual overload,' Cernan wrote. 'I had poked my head inside a kaleidoscope, where shapes and colors shifted a thousand times a second . . . This was like sitting on God's front porch.'

Unfortunately, also in common with Leonov, Cernan's walk

rapidly turned into a battle, and for the same reason. His pressurised spacesuit was so stiff that he likened it to 'hardened plaster of Paris, from fingertip to toe', and also to 'a rusty suit of armour' that 'took on a life of its own'. All movement was a giant, energy-sapping struggle. The suit's cooling system struggled to cope with the heat from Cernan's exertions and his face mask misted over.

The 25-foot umbilical looped around him and resisted his attempts to haul on it. 'I felt as if I was wrestling an octopus,' Cernan said. He, like Leonov, became a doll on a string, 'absolutely helpless', tumbling and turning with almost no control over his destiny. This went on for thirty minutes before he managed to snatch at a handrail on the craft and 'pawed my way back to the open hatch, a drowning man reaching for a titanium beach'.

But his struggles were really only beginning. Now he had to test the Astronaut Maneuvering Unit, which was essentially a precociously early version of the MMU jet-pack eventually used by Bruce McCandless. Cernan had trialled this large and cumbersome-seeming unit as best he could on the ground and in bursts of zero gravity on the 'Vomit Comet', clad in steel-mesh trousers to avoid scalding his backside with jets of hot gas. (His spacesuit included special pants of woven nickel-alloy to provide the same protection – think Wallace in *The Wrong Trousers*.) But now the AMU would get its ultimate – and surely premature – test, in space.

The AMU was stored at the rear of the craft. Pausing only briefly to recover his strength, Cernan set off to fetch it. It was now orbital night, and he could barely see what he was doing. He struggled to unfold the AMU's arms and back his body into it. His spacesuit overheated, his heart-rate topped 180bpm, and his helmet's visor, instead of merely misting over, now properly fogged up on the inside. By stretching his neck to its full extension, Cernan could just use the tip of his nose to clear a small patch of the visor to see through.

The most ruinous development, though, came when he swapped the spacecraft's umbilical for the AMU's tether and promptly lost radio contact with Stafford inside the capsule. Flying blind would have been one thing, but flying deaf, too . . . At long last, a bitterly disappointed Cernan conceded the need to call the AMU activity off. He reconnected to the spacecraft umbilical and gingerly felt his way back round the craft to the hatch.

Again the shadow of Leonov was waiting for him. Cernan's stiff, ballooned spacesuit made it difficult for him to wedge himself back into the Gemini far enough to be able to close the hatch. Like Leonov, he was obliged to endure an undignified and painful bout of squeezing and wedging. In fact, folding his cramping body into the space and getting his helmet down below ceiling level was agony. Finally, though, the hatch was shut and Cernan's trial was over. His spacewalk had lasted two hours and seven minutes.

Alas, on this occasion there would be no photographs to show for his trouble. Cernan accidentally kicked the mounted Hasselblad camera on his way in, dislodged it, grasped at it, missed, and sadly watched all his pictures float away into space, never to be developed.

Stafford helped him depressurise the suit, which finally softened around Cernan's aching limbs. When he removed his helmet, Stafford was alarmed at the sight of his colleague's almost luminously red face. He took the water gun, used for drinking and for rehydrating food, and squirted cold liquid at Cernan's head. To do so was to risk sending globules of water floating off into the capsule, possibly getting into the electrics and causing short circuits. But this was an emergency. Cernan cooled down and the circuitry survived.

After a much-needed rest, Stafford and Cernan prepared to return to Earth. But Gemini 9 still had one cruel trick to play. All went smoothly through re-entry, but when the capsule splashed down, it bumped hard against the wall of a wave, jolting them both violently and making a loud bang. Suddenly, to their dismay, water started washing into the capsule. The impact, they thought, must have cracked the capsule and now it was filling up around them.

They could, of course, sling open the hatch. But it would only take one big wave to come pouring into the capsule and they would be sent to the bottom of the sea even faster. They held tight and chose instead to stare down anxiously at the water around their ankles. To their immense relief, it did not appear to be rising. So maybe this situation wasn't quite as dire as it seemed.

In fact, the impact of their splashdown had cracked an internal water pipe. What Stafford and Cernan were standing in was the last of their drinking water. The capsule would still be afloat when the rescue helicopter and the frogmen reached them.

In weight-loss terms, Cernan had outdone Leonov: he had shed 13.5lb across the three days of the mission. He also had a patch of what appeared to be sunburn at the base of his spine, which the medical staff were mystified by. Had the Sun penetrated his suit? Or was it something to do with his rogue struggles with the AMU? This anomaly seems not to have been explained.

Years later, Cernan could still feel tired just recalling that spacewalk. He returned with a sense of failure, worried less about what the world would make of it than what his fellow astronauts would think – knowing that, ironically, among the people better placed than any others to understand what he'd been up against and how over-ambitious the mission had been, there would nevertheless be whispers about how this rookie had gone and screwed up the job.

In fact, Cernan would be personally exculpated by the post-mortems on the mission, which showed very clearly how the odds had been stacked against him. It also helped his reputation when the next two NASA spacewalkers didn't fare much better. Michael Collins on Gemini 10 was sent out on a tether to walk from his capsule to the nearby Agena target vehicle and retrieve an experiment package. The problem was, nobody had really thought about what Collins might hold on to on the Agena while he did this. He grabbed at the craft's entirely smooth docking collar with his cumbersome glove, slipped off it and – as he put it later – 'went cartwheeling ass over tea kettle' until he reached the end of his tether, at which point he found himself swinging out in space in a giant arc. Collins squeezed at the hand-held thruster unit – 'this dorky little gun' – to get himself under control. Then he worked his way back up to the Agena. On his second attempt, he steadied himself by reaching inside the docking collar and grabbing a fistful of wires – probably less than ideal – and used his other hand to detach the package.

Mission accomplished, then, but with very few points for style, and with some unhelpful elements of extreme jeopardy.

'It was not clean analytical engineering,' Collins reported many years later with typical dryness. 'It was more acrobatics and "guy on a trapeze" stuff that you don't think about in the space programme.'

On the same mission, Collins would discover that even being a humble, non-acrobatic 'standee' had its grim perils. In a separate EVA, Collins was required simply to stand up in the Gemini's open hatch and perform some measurements of the stars. Suddenly he was blinded by stinging tears, possibly the result of granular lithium hydroxide escaping from the suit's carbon dioxide purging system and acting as an irritant. Whatever it was, it was impossible for Collins to keep his eyes open, and it left him attempting to perform the complicated and utterly essential task of getting the capsule hatch closed behind him almost entirely by feel. He struggled for a couple of minutes until his eyes abruptly cleared.

Collins was once asked if that experience had been frightening. 'Yes,' he replied – a rare instance of an astronaut giving an unqualified affirmative answer to that question.

Next up was Dick Gordon in Gemini 11, who had an even harder time than Collins. On an EVA intended to last two hours, Gordon was to retrieve a 30-metre-long tether from the Agena and attach it to the Gemini. This he achieved, but only at the expense of utterly draining himself through over-exertion. His heart-rate soared, he was covered in sweat and he temporarily lost vision in one eye. Ground Control advised him to rest up mid-walk and just hang there for a while, at which point Gordon became the first human to take a management-sanctioned break in the void. But even after a spell of floating peacefully he was still too shattered to continue and went back in.*

The cumulative effect of these early experiences led to fundamental changes to the training for future spacewalks, and also to their parameters. As astronaut after astronaut returned to the capsule red-faced and sore, NASA were obliged to address, as one of their post-mission reports put it, 'the difficulty of physical labour in a hard-suit environment'. Work started in Houston on a giant pool in which spacewalks could be accurately drilled and pre-tested in underwater simulations. The importance of handles, rails and footplates had now been categorically demonstrated. Also the need for more flexible suits with better cooling systems.

The idea of manoeuvring units, hand-held or otherwise, was shelved for a while – indeed until Bruce McCandless briefly

---

* Both Michael Collins and Dick Gordon went on to pilot lunar orbiting modules, on Apollos 11 and 12 respectively.

revived it in triumph in 1984. Instead, the concentration for a while was on far simpler EVAs, testing the possibility of doing fairly basic chores in space: tearing Velcro strips, operating simple tools such as wrenches. On Gemini 12 in November 1966, Buzz Aldrin successfully completed all his jobs and, logging five hours in three separate walks in a greatly modified suit, and with hand-holds and work-stations to assist him, was able to return inside comfortably, with his eyesight intact and not even out of breath.

'So what was the problem with spacewalking?' Aldrin's demeanour afterwards seemed to suggest. The other astronauts muttered to each other about how much easier Aldrin had had it, and how he had merely been (in the words of Cernan) 'working a monkey board'.

Aldrin ignored them. The shuffle of the mission order that had arisen from the deaths of See and Bassett had a significant ripple effect for him. It moved him along the line to back-up crew for Apollo 8, which in turn put him in line for a place in the crew on Apollo 11.

And Apollo 11, of course, would find him playing a part in unarguably the greatest EVA of them all.

V. ON THE LADDER TO IMMORTALITY

As we have seen, nothing much fazed Neil Armstrong while at the controls of a spacecraft. And it's a good job too, because, on the final descent to the Moon in the Apollo 11 Lunar Landing Module, many things tried to.

Apollo 11 had launched on 16 July 1969 and set off on its four-day journey to the Moon. On the first morning, Bruce McCandless, as Capcom in Houston, had woken the crew and, after a few pieces of flight-related business, read them their morning news digest. 'Europe is Moon-struck by the Apollo 11 mission,' he said, reading off the Associated Press wire-feed. 'Newspapers throughout the continent fill their pages with pictures of the Saturn V rocket blasting off to forge Earth's first link with its natural satellite. And the headline writers taxed their imagination for words to hail the feat. "The greatest adventure in the history of humanity has started," declared the French newspaper *Le Figaro*, which devoted four pages to reports from Cape Kennedy and diagrams of the mission. The tabloid *Paris Jour* proclaimed, "The whole world tells them bravo."'

McCandless had also thrown in another, unrelated news item for the crew's delectation. 'London UPI: The House of Lords was assured Wednesday that a midget American submarine would not "damage or assault" the Loch Ness monster. Lord Nomay said he wanted to be sure anyone operating a submarine in the Loch "would not subject any creatures that might inhabit it to damage or assault". He asked that the submarine's plan to take a tissue sample with a retrievable dart from any monster it finds can be done without damage and disturbance. He was told it was impossible to say if the 1876 Cruelty to Animals Act would be violated unless and until the monster was found. Over.'

'Roger,' said Armstrong. 'Thank you, Bruce.'

And that was the news.*

At a hundred hours and twelve minutes into the flight, Eagle, the peculiar, fragile-looking and oddly insect-like module charged with bearing Armstrong and Buzz Aldrin to the lunar surface,

---

* Nessie hunters sought permission to use a sonar-equipped submarine vessel in their searches. The Joint Parliamentary Under-Secretary of State for Scotland, Lord Hughes, assured the House that 'the organiser of the Loch Ness Phenomena Investigation Bureau Limited has given assurances to the Chief Constable of the Inverness Constabulary that the submarine operations have no aggressive intent'. Nessie continued to be elusive.

separated from Columbia, the Command Module piloted by Michael Collins, who then flew around the lander to give it one final visual check. And then, after another hour of solo orbiting, the Lunar Module fired its engine to give retrograde thrust and the descent began.

Eagle required Armstrong and Aldrin to stand side by side, held to the floor by harnesses. At first it flew them face down, so that the Moon loomed directly up at them from 50,000 feet through the tiny triangular windows, getting larger, growing closer.

Noticing that the landmark checkpoints were appearing in his window two seconds ahead of time, Armstrong could calculate that Eagle was going to arrive about 2 miles beyond the chosen landing point. But this didn't really worry him. After all, as he told James R. Hansen, the author of *First Man*, 'There wasn't going to be any welcoming committee there anyway.' In any case, he assumed the computer, programmed to land automatically, would eventually compensate and put them right.

At 46,000 feet, the module turned onto its back to engage its landing radar, flipping Armstrong and Aldrin onto their backs, too. Now, where the Moon had been in their windows, they could see the Earth, a disc in the sky.

Down they continued, staring upwards, trusting the machinery. At 40,000 feet the landing radar engaged, its information feeding the computer which in turn sent little blips of power to the jets, correcting the module's direction, as Armstrong had known it would. But was it enough? Or would they still come in too fast and overshoot?

At that point, the emergency alarm sounded, accompanied by a flashing red light on the console and, on Aldrin's computer screen, an error code: 1202.

Neither Armstrong nor Aldrin knew what a 1202 error code was. It had never come up in simulations. Flight Director Gene

Kranz, at Mission Control in Houston, didn't seem to know either. Indeed, everyone on the ground seemed baffled by it. It definitely wasn't one of the big ones – one of the emergencies requiring that they immediately abort the launch and send the LLM back up to rejoin Michael Collins, orbiting the Moon in the Command Module. The most informed view seemed to be that it was just a glitch – the computer getting overloaded and resetting itself.

So, an alarm, yes. But an intermittent one, not a continuous one. Nothing to be unduly . . . alarmed about.

'We're GO on that alarm,' was the word passed on to the module by the astronaut Charlie Duke, now on shift as Capcom.

Twice more that alarm sounded as Eagle continued its descent, and twice more the instruction was to ignore it.

At 7,500 feet, the module slowed and tipped from its face-up position so that Armstrong and Aldrin were now descending vertically. Now it was the Sea of Tranquility that loomed in their windows.

'Eagle, Houston. You're GO for landing,' said Duke.

'Roger, understand,' said Aldrin. 'GO for landing, 3,000 feet.'

Almost at the same time, the alarm went off again: 1201 this time, not 1202. Which made a change. But up came the same instruction from the ground: ignore it.

'We're GO. Same type. We're GO.'

Armstrong began to look ahead to the computer's estimated landing spot. It looked OK.

Another alarm. Aldrin cleared it. The alarm immediately sounded again. Aldrin cleared it again.

Now they were at 1,000 feet and Armstrong was having to revise his initial assessment. It didn't look OK at all down there. Ahead of them lay not the smooth, flat surface they had planned for but a huge crater and a patch of ground beside it strewn with rocks and large boulders. Crash down onto one of those boulders and the fragile module, whose walls were in places no thicker than a Coke can, could easily split. Alternatively, set down on a badly uneven patch of ground, perhaps tilted into a crater, and the module could find itself impeded from launching directly upwards when it was time to leave. In either of those scenarios there would be no coming home.

What Armstrong realised he needed to do was ignore the repeating alarms, take manual control of the module, fire the engines to hold the descent at 350 feet, and fly on in search of somewhere better to land.

He disengaged the autopilot, selected ATTITUDE HOLD and sent Eagle forward.

This was the point at which the operation to get the lander down safely necessarily ceased to be a collaboration between Houston, its computers and Eagle. It was now all on Armstrong.

Oh, and just one other thing: they were low on fuel.

Back on Earth, thought had inevitably been given to the possibility that this huge public spectacle, the culmination of years of anticipation, which seemed to have brought most of America to its television sets and radios, might very possibly end badly. Richard Nixon's speechwriter, Bill Safire, had readied the following words for the President to read to the country and the watching world in the event that the first astronauts to reach the Moon became stranded there:

Fate has ordained that the men who went to the Moon to explore in peace will stay on the Moon to rest in peace.

These brave men, Neil Armstrong and Buzz Aldrin, know that there is no hope for their recovery. But they also know that there is hope for mankind in their sacrifice.

These two men are laying down their lives in mankind's most noble goal: the search for truth and understanding . . .

In their exploration, they stirred the people of the world to feel as one; in their sacrifice, they bind more tightly the brotherhood of man.

In ancient days, men looked at stars and saw their heroes in the constellations. In modern times, we do much the same, but our heroes are epic men of flesh and blood.

Others will follow, and surely find their way home. Man's search will not be denied. But these men were the first, and they will remain the foremost in our hearts.

For every human being who looks up at the Moon in the nights to come will know that there is some corner of another world that is forever mankind.

Fortunately, with the fate of his spacecraft, his colleague's life and America's Moon mission transferred to his own two hands, Armstrong would now pull off a piece of unprecedented and utterly nerveless piloting to ensure that, among other things, Nixon's touching obituary of him and Aldrin remained in a drawer.

Picking out an open patch up ahead unpopulated by boulders, Armstrong flew towards it. Setting the module down squarely was going to be crucial. A ragged landing could snap off one of its thin and brittle legs and, like arriving in a crater, leave the craft in a position from which it might never successfully emerge.

One problem, though: the dust. As they got closer to the surface, Eagle began to kick up clouds of the stuff which seemed instantly to congeal into a shifting grey blanket below them, blurring out all the details of the surface with the exception of the larger rocks that were tall enough to poke their heads above it.

Armstrong was seeing the terrain ahead of him and then watching it disappear in a flurry of dust as he closed in on it. He was basically being required to land from memory.

His pulse rate rose to 150.

Another light lit up on the console: DESCENT QTY.

It was the fuel gauge. The fuel was low, lower than it had ever been on any of their simulations.

'Quantity light,' reported Aldrin.

In Houston, that call from Aldrin effectively started the clock on a 'bingo' fuel call: after ninety-four seconds, they were going to have to call 'bingo'. If Armstrong didn't think he could get the craft down in the next twenty seconds after 'bingo', he would have to abort.

Armstrong dropped the module vertically into the dust cloud. As it slowly lowered to the ground, the module seemed to want to move backwards. Armstrong arrested it with the throttle. Then it seemed to want to move sideways. Armstrong arrested it again.

'Thirty seconds,' said Duke.

'Forward drift?' said Armstrong.

'Yes,' said Aldrin.

Armstrong applied the throttle. The dust flared around them even more thickly. It was like dropping into fog.

'Contact light,' announced Aldrin.

The lander touched down so gently onto the Moon dust that neither of the astronauts felt it happen.

'OK. Engine stop,' said Aldrin.

They were on the surface with only seconds of fuel remaining before an abort call.

'We copy you down, Eagle,' said Duke as the Mission Control room began to explode with relief around him.

'Houston, Tranquility Base here,' said Armstrong. 'The Eagle has landed.'

As it happened, the fuel gauge had been playing tricks and introducing a little extra drama where none was particularly needed: it was later established that Armstrong could have flown on his remaining supplies for about another minute had he needed to.

But, of course, he wasn't to know that as the craft descended and the rocks loomed and the alarms buzzed and the dust flew and the fuel ran low and the eyes and ears of the entire world strained for news of his progress.

What he did know was that he was now on the surface of the Moon. Through the window above him he could see the Earth again, 250,000 miles away.

It was time to get out and explore.

Except not quite. With little thought for the dramatic shape of this world-changing moment, NASA had decreed that the first two things the astronauts should do after landing on the Moon were (1) perform a plugs-out launch run-through, and (2) sleep. The first exit from the module was accordingly set for ten hours after engine shutdown.

Ten hours!

However, the astronauts, NASA quickly realised – and this was perhaps predictable in the circumstances – were far too wired to get much sleep.* The EVA was brought forward four hours.

So, six hours after touching down, on 20 July 1969 – a Sunday afternoon on the east coast of America – black and white images began to emerge from the remote camera attached to the Lunar Lander and flicker onto televisions around the world; images so fuzzy and indeterminate at first that Armstrong's six-year-old son, Mark, stared at the screen in their sitting room and asked, in confusion, 'Why can't I see him?'

But then Armstrong was at the bottom of the ladder and taking his 'one small step for man, one giant leap for mankind', and

---

* All of the Apollo crews would experience frustration with the requirement that they spend precious hours on the Moon asleep. Gene Cernan, on Apollo 17, remembered being in bed and thinking, 'I got on top of that Saturn V to get here and now I'm lying around inside this LM. Why am I sleeping at all?'

everybody could see him well enough – and would continue to see him forever.*

There had, of course, been huge debate around the subject of who was going to be first out of the LM – Aldrin or Armstrong – and *why* they were going to be first out, and that debate would rumble on loudly afterwards. And it came down ultimately to the shape of the craft, to the interior layout and the shape of the hatch. Simulations had categorically demonstrated that it was less risky to have Armstrong go out first than to have Aldrin climb past him in order to get out.

Or that's what NASA said at the time. When Chris Kraft, the NASA Flight Director and later Director of the Manned Space Center, published his autobiography, in 2001, it became apparent that Armstrong went down the ladder first because key people at NASA wanted it to be that way. Kraft knew in advance what the first person who walked on the Moon would become and remain for the rest of his life. He knew what would be waiting for that person on the other side of this adventure, how they would be instantly transformed by the alchemy of fame and attention. And Kraft, and many others whose view counted, simply thought Armstrong was better suited to that onerous role than Aldrin.

Relentlessly focused and with an apparently iron-cast determination, Aldrin allowed more of himself to come to the surface than Armstrong ever did. He was obsessive about details, mission-driven to an exceptional degree even among astronauts. On the

* On the much-discussed missing 'a' from Armstrong's famous declaration (not 'one small step for man' but 'one small step for a man' would seem to have been the perfect phrasing, contrasting with 'mankind' in the second half) there are three main theories: (1) that Armstrong screwed it up, (2) that his Ohio accent swallowed the 'a' inside the 'r' of the 'for', and (3) that a radio communications glitch cut it. I tend to favour theory two while reserving some credence for theory three.

day of his selection to be part of the single greatest adventure in the history of human exploration, he went home to tell his wife. Joan Aldrin later reported that Buzz spent that evening explaining to her at length the various means that were being developed for collecting rocks from the Moon.*

Altogether, the chemistry of the Apollo 11 crew seems to have been unique, and not in the way we perhaps might expect. There were strong bonds of some kind in every other Apollo crew, yet not in this one, the most famous of them all. Michael Collins, the pilot, described their team as 'amiable strangers'. This was a crew that arrived everywhere in three separate cars, and went their different ways at the end of the working day; a crew that functioned perfectly professionally, with barely a cross word, but took their relationship no further.

Accordingly, when it became clear that he wasn't going to get the job of first person out, a role he had never concealed his desire for, Aldrin seems to have felt no qualms about taking a professional decision and working hard behind the scenes to get the management at NASA to change their minds – which he could not.

Meanwhile Armstrong, as ever, couldn't see what the fuss was about. Why should anyone care which of two astronauts went down a ladder first? Wasn't landing on the Moon the real achievement here? Standing on it was just an encore after the main show, which was getting a craft containing two humans safely to the lunar surface. Logically, to Armstrong's mind, it made zero difference whether it was 10 feet of aluminium leg between the astronauts and the dust, or the thickness of the soles of their boots:

---

* I can relate that I once took a phone call in an airport from Buzz Aldrin, who wanted to discuss a spaceflight-related project which had been mooted at a conference we had both attended. I listened near Duty Free while he talked for twenty minutes. Meticulous does not begin to describe him.

both Armstrong and Aldrin had reached the Moon in unison the moment the Lunar Module touched down, and so the rest of it was entirely beside the point. It simply didn't matter.

But it did matter, didn't it? It mattered that a foot was set down on the ground. The symbolism of the footprint on the surface – that was what the world was going to take away from this moment, not the planted leg print of the Eagle lander. Until Armstrong had descended that ladder and set his boot down in the dirt, the Moon had not been *walked upon*.

And this was what it all boiled down to in the end, even after the unprecedented feats of rocketry and all the extraordinary technical innovation which had safely flown two astronauts to a destination a quarter of a million miles from home: the basic human act of walking. The whole history of human spaceflight had been tending determinedly towards this moment, and ultimately, when it arrived, it wasn't going to be about machines and what machines did, it was going to be about humans and what humans did.

So, yes, I think we can categorically say that the first step mattered, and that who took that first step mattered too, and Armstrong was plainly wrong.

Yet we could also argue this: the fact that he was someone who didn't really believe it mattered also meant that Armstrong was exactly the right human for the job.

VI. SOLDERING IRONS AND SIMPLE MATHS

In October 2019, NASA astronauts Christina Koch and Jessica Meir clambered around the outside of the International Space Station using handrails and metal carabiners, pulled out a failed power distribution unit – part of the battery system that distributes solar power to the station – inserted a replacement and then lugged the faulty one back inside the airlock so that it could be returned to Earth and scrutinised. The operation lasted seven hours and seventeen minutes and meant they had just completed the first all-woman spacewalk. This was clearly a significant moment in spaceflight history, although the NASA astronaut Tracy Caldwell Dyson put the trumpets in reasonable context when she remarked, 'As much as it's worth celebrating, many of us are looking forward to it just being normal.'*

Afterwards, President Donald Trump called the space station to congratulate Koch and Meir on what they had just achieved. During their conversation, Trump formally heralded the occasion as 'the first time for a woman outside of the space station'.

Well, not exactly. Koch alone was on her fourth spacewalk and Meir had just become the eleventh woman ever to do one. All in all, it was the thirty-sixth time women had been outside in space, with Peggy Whitson alone being responsible for ten of those EVAs. In other words, had he but known it, there had been getting on for as many spacewalks involving women as there had been Presidents before Trump.

And foremost among those walks had been Svetlana Savitskaya's. In 1984, thirty-five years before Koch and Meir climbed out together, Savitskaya became the first woman to haul herself out into the void, an act to which you might say she had been building all her life: by the time she was seventeen, Savitskaya had completed 450 parachute jumps and would go on to hold world records for stratosphere jumping, flinging herself out of planes from heights in excess of 14,000 metres.

The spacewalk came during her second mission to the Salyut space station. (Her first mission there, in 1982, created the first

---

* Increasing the progress towards that normality, Koch and Meir's 2013 NASA astronaut candidates class was 50 per cent women.

mixed-gender crew on a space station.) She was tasked with testing out the Universalny Rabochy Instrument, or Universal Working Tool, a device created in Kiev and designed to make it possible to cut, solder, weld and braze in space. Think Dr Who's sonic screwdriver, but with more practical applications for zero-gravity metalwork.

'I didn't understand the point of it,' Savitskaya later said of the strictly academic tasks she had been set. She thought it had merely exposed her to the risk of burning a hole in the space station or setting fire to her spacesuit. Still, out there on her EVA, Savitskaya dutifully sawed away at various trial slabs of titanium and stainless steel, and performed miscellaneous pieces of tin and lead soldering. It was mind-bogglingly dangerous: even today's superior spacesuits could only maintain pressure with the tiniest of holes in them – 6mm or less. Anything larger than that, game over. Yet Savitskaya exposed herself to the risk, and the sense of what was possible in space got nudged along once again.

Now come forward a few years, to late 1993, and the Space Shuttle *Endeavour* mission STS-61. This is the first Hubble Space Telescope rescue operation, which will see *Endeavour* fly to 370 miles above Earth, come to within 35 feet of the telescope, capture it with a robotic arm and park its 43.5-foot body in the Shuttle's payload bay for an overhaul.

In a series of spacewalks over eleven days, Story Musgrave, Jeff Hoffman, Kathryn Thornton and Tom Akers, working in pairs, performed five back-to-back EVAs on that mission, spending almost thirty-five and a half hours outside the Shuttle, fixing the telescope's gyroscopes, control units and electrical circuits and beginning to rescue the Hubble from the standing joke status to which it had descended. (You know your $1.5 billion hardware is in trouble when it features as a gag in one of the *Naked Gun* movies.)*

The NASA astronaut Mike Massimino was part of a later Hubble mission – STS-109 on Space Shuttle *Columbia* in 2002. It was his first spaceflight – and quite a way to make your debut. From the International Space Station you can't see the Earth whole, but from higher up, at the Hubble, you could – the entire marble, 'the most magnificent and incredible thing you've ever seen in your life', Massimino said.

In partnership with James Newman, and with that magnificent marble in the background, Massimino conducted two walks totalling fourteen and a half hours – two full days' work in space – leaving the Hubble with new solar arrays and a new power unit, among other upgrades. This was how far EVAs had come in the course of half a century, from Leonov's madcap hatch escapade, via Savitskaya's experimental metalwork, to whole working days of hi-tech electrical engineering in the void.

We've learned how to stabilise ourselves on a spacewalk with footplates, handrails, clamps and tethers. We use robotic arms to move around with ease and operate complex tools – no more flailing around tugging on umbilicals or pointing futile zip-guns. Our spacesuits fit much better; at the last count NASA offered over forty different sizes of glove. But some aspects of spacewalking remain the same. Astronauts still return utterly exhausted, often red-faced and drenched in sweat. Occasionally tools still get lost,

---

* It's pictured on the wall of a nightclub alongside the *Titanic* and the *Hindenburg*. Chief among the telescope's many, now mostly forgotten, teething problems was a primary mirror which wouldn't focus because its edges had been polished too flat – or, to employ the magnificent euphemism devised by a highly embarrassed NASA for this fault, a mirror which had suffered 'a spherical aberration'. Thanks to the Shuttle and its repair crews, the Hubble has gone on to be a game-changer in terms of our knowledge of the visible universe.

floating off into space 'on a definite trajectory going somewhere', in the words of Ed White. Spacewalking remains the most physically and mentally demanding task for any astronaut, and the one that carries the greatest risk.

Massimino once described spacewalking as like being sent out to pitch in the baseball World Series, in front of a full house, and having done loads and loads of practice – but never actually having played a baseball game before. It's as good a description as any of the anticipatory feelings ahead of a spacewalk that I've come across. In fact, it applies to so many experiences in space. You rehearse and rehearse but it's all theoretical until you get out there.

But out there, in the case of spacewalking, means out there in the perfect white sunlight, which, as Massimino noted, is entirely unlike sunlight as we see it on Earth: 'It was like I was seeing things in their purest form, like I was seeing true colour for the first time.' And though it's not given to many of us to zip around the place like Bruce McCandless, it's still the most extraordinary thing to find yourself, as I did in January 2016, sliding feet-first from the airlock, and venturing out into space.

The preparation that brings you to that moment is as thorough as any you will have undertaken on your astronaut journey. You will have prepped your spacewalk on the ground, you will have prepped it in the pool, you will have prepped it with a Virtual Reality headset on.* You will have been visualising every shift, every translation, running them through your mind ceaselessly in

---

* A proper, cutting-edge Virtual Reality headset, not the system rigged up for my training which involved a pair of goggles and a laptop turned upside down and strapped to my head. No points for style, obviously, but decent enough marks for fidelity.

the days leading up to it. You will have every move down like a dance routine.

And then it's time to go, and you're stepping into your spacesuit – a job which weightlessness, and having nothing to push your feet against, makes additionally complicated. During my suit-fit check, a couple of days before the walk, I actually pinched a nerve in my shoulder while wrestling my way into the suit, and one side of my left forearm and my left thumb turned numb for a little while. My heart sank: to be ruled out by injury at that stage, and for that reason . . . But I could still grip, so the flight surgeon gave me the go-ahead. Take care getting into that suit, is my advice.

You'll be in that stiff, restrictive armour for a few hours before walking, breathing in oxygen and flushing the nitrogen from your system. Then you'll pass into the airlock and seal yourself in. It's very cramped in there, by the way – stuffed full of gear and baggage and with enough room for two suited astronauts, though only if they go head to tail. So you'll be lying there with your crewmate's boots in your visor while the airlock depressurises.

But then the hatch gets opened and that pure white sunlight floods in – a sensational moment. And you'll wait for your crewmate to get out and get secure, then you'll pass out to them the equipment you'll be needing, taking care not to get any tethers tangled. And then you'll join them in space.

And if you're lucky, you'll get a moment to just let yourself float on the short tether connecting you to the station, its structure so sharply delineated in the Sun, and allow yourself to feel the serenity of it all.

You're weightless, no forces exerting themselves anywhere on your body. The temperature, regulated by the softly humming ventilator in your suit, is somehow perfect, and it's just you, with the Earth spinning below you, and the space station right ahead, and, just over your shoulder, the universe.

And you may feel two things at that point, both at the same time. Firstly, that you really shouldn't be here; that this entirely hostile environment was specifically designed to exclude you, a humble, fragile human being, and that the almost overwhelming experience you are having, of floating in the black void and looking down on the Earth, was not one that any human was intended to have.

And secondly, that this is the most tranquil and natural thing that you have ever done.

But of course, there's a job to do, so you'd better stop floating and get on with it. Those Sequential Shunt Units don't replace themselves, you know. They're the size of a small fridge, too – but, of course, that's where weightlessness is your friend. So it's time to make your way out to the far end of the ISS's golden solar array, where you're going to whip out the old unit and plug in the new. Those SSUs have proved tricky to unbolt in the past, but if you're lucky they'll slide in and out like a dream and you'll be heading back for the airlock with the old one before you know it.

There'll be hand-holds a-plenty along the way, of course, but do take special care while traversing the CETA spur, which is basically a short stretch of aluminium tube which you'll have to pull yourself along with your feet dangling. You'll basically be hanging from the ISS at that point (albeit with the additional comfort of your tethering). If you do what I did, and look down at that moment, you may see, 400km below your boots, Western Australia going by. And that may cause you, momentarily, an urge to grip tightly, your brain reverting to the Earthbound assumption that if you let go, you'll fall. Self-preservation is a powerful instinct.

Best cure for vertigo? Wiggle your toes. Chris Cassidy, a NASA astronaut and former Navy SEAL, taught me that. And it worked for me: the moment passed and I carried on along the tube and back en route to the airlock.

And afterwards, back inside, with the job successfully completed, having a cup of tea and putting your feet up, there will come – if your experience is anything like mine – a sense of elation unparalleled in astronaut work, one that can almost make you sad that at some point you're going to have to go home.

CHAPTER SEVEN

GETTING HOME

'You get a feeling that people think of these men as not just superior men but different creatures. They are like people who have gone into another world and returned, and you sense that they bear secrets that we will never entirely know.'

– Walter Cronkite, CBS News, commenting on the return of Apollo 11

I. LOST AND FOUND

April 1970. The television presenter Cliff Michelmore looks gravely into the camera and tells viewers of the BBC: 'The best thing we can do now is just listen and hope.'

The BBC is broadcasting live black and white pictures beamed from the deck of USS *Iwo Jima*, waiting hopefully in the Pacific for the appearance in a cloudy sky of Apollo 13's Command Module Odyssey (after Homer's epic narrative – no space capsule was ever more aptly named). And it is intercutting this footage with equally black and white pictures of Michelmore, science historian and author James Burke, astronomer Patrick Moore and a number of others hopefully watching those live images in a studio in London.

In other words, fully forty-three years before Channel 4's *Gogglebox* and a good twenty-one years before Sky's *Gillette Soccer Saturday*, the BBC are categorically demonstrating the possibility of creating oddly gripping television out of pictures of people watching television.

We are in the communications black-out period – usually around four minutes of agonising suspension, but in this case, because of Apollo 13's more shallow angle of entry, extended to six minutes while the capsule burns through the layers of the atmosphere encased in a ball of plasma, slowing its return from the Moon at over 23,000mph, silent to us on Earth and completely out of sight.

And at the end of those six, heavy minutes, Houston will attempt to contact Apollo 13 and find out if its heat shield has in fact held out during the trauma of that period, and if indeed the three crew members of Apollo 13 are still alive and coming home.

That's the Apollo 13 crew – Jim Lovell, John 'Jack' Swigert and Fred Haise – whose Service Module has suffered a ruptured oxygen tank (Lovell: 'Houston, we've had a problem'); whose mission to the Moon has been abandoned after two days to become, instead, a frantically improvised battle to get back, ground and space coordinating as never before; who have camped, all three of them, for days in the cold, dark, damp, two-person Lunar Module to save power; who have literally been using duct tape to stick bits of plastic to the cardboard covers torn from the craft's instruction manuals just to maintain air-flows and survive; who have somehow made it all the way back from lunar orbit, and who even now may suffer a heat-shield failure on the way through, rendering all those incredible and unprecedented efforts in vain.

Public interest in the Moon missions might have dipped slightly in the immediate wake of Apollo 11's triumphant lunar landing the previous summer, but it's back now. Britain is tuned in, the world is tuned in. Burke has his fingers not so much crossed as twisted together. Michelmore is chewing down hard on a fingernail. The pictures show black helicopters hovering distantly over a grey sea.

Finally, the Capcom, Joe Kerwin, makes the call.

'Odyssey, Houston standing by. Over.'

There's a short, agonising pause. And then from the sky the voice of Jack Swigert crackles through: 'OK, Joe.'

The BBC studio dissolves into smiles and giant sighs and bouts of most un-BBC-like shoulder-squeezing.

DROGUES DEPLOYED flashes on the screen.

The camera pans the grey and white sky until it finds the merest glimpse of a darker grey dot, loses it, and then finds it again.

'There they are! There they are!' shouts Burke, elated.

Note: not 'There it is' but 'There they are', this story long since having ceased to be about the safe delivery of a spacecraft and become entirely about the safe delivery of the humans inside it.

'They've made it!' shouts Burke as applause and cheers break out on the soundtrack. 'And listen to the people on the boat!'

Actually the applause we can hear seems to be exploding out of Mission Control where the greatest and least plausible rescue act in spaceflight history has somehow come out right, and where a vast release of tension can now take place. Gene Kranz thought returning the Apollo 13 crew alive was 'NASA's finest hour', eclipsing even the Moon landing. It would be hard to argue with him. Certainly Jim Lovell's disappointment at having to abandon his descent to the lunar surface gave way eventually to a sense that he had been part of something at least as big.*

Mind you, there's still time for one last heart-stopping moment for viewers as what looks like a puff of smoke shoots from the descending capsule. But it's quickly explained, by Houston and

* If Tom Hanks plays you in the movie of your life, the chances are you did OK.

Burke almost in unison, as vented fuel and nothing to be worried about.

And then the capsule drops into the waves and the parachutes flop into the water and the crew really are home.

'They're in!' cries Burke. 'And I make it no more than five seconds late! No more than *five seconds late*!'

The entire BBC team is rocking backwards in its seats exultantly and slapping the desk.

Few returns pack emotional heft like the return to Earth of an astronaut crew – and not just Apollo 13's crew but the crew of any returning spacecraft. What guarantees tension every time is the fact the plot will always include that vital cliff-hanger: those minutes of radio silence, that agonising disappearance at the crucial moment. It would seem contrived, almost too much, if someone wrote it, if it wasn't naturally occurring, the product of science – plasma's gift to drama.

There had been a similar moment with Apollo 8 a year and a half prior to Apollo 13, albeit at a different stage of the return, and with distance rather than plasma creating the dramatic loss of contact. That crew achieved their unprecedented orbit of the Moon, passing just 69 miles above the lunar surface, but in order to return to Earth safely they had to rely for the first time on a successful Trans Earth Injection – a propulsive firing of the engine which, as if the Hollywood directors had got their hands on the flight plan in advance, would have to take place with the spacecraft on the Moon's far side, beyond contact.

It was approaching midnight on Christmas Eve (those Hollywood directors again, surely). If the burn was successful, the crew would re-establish radio communications about twenty minutes later, at nineteen minutes after midnight. If time went on and there was no contact, the burn had clearly failed.

So, on that occasion too there was just waiting and silence and held breath.

And that time too the radio crackled and Jim Lovell, who was on that mission as well, said, 'Please be informed there is a Santa Claus.'

But this is the gut-punching script that spaceflight's return journey automatically writes for us every time a spacecraft comes back to Earth. The drama of re-entry insists that, at the moment of maximum tension and uncertainty, when everything is most critically at stake and the crew at their most vulnerable, the capsule will disappear entirely off our radars for a while, leaving those of us on Earth with no option but to cross our fingers and hope.

It's almost as though you have to find out what it feels like to lose them before you can properly get them back.

II. WALTZING HOME

How many of the pioneers whose journeys we have been following in these pages nearly didn't make it through their mission's inevitably fraught closing stages? How many of them almost fell hard right at the end?

So many that you could easily begin to wonder: is there any such thing in spaceflight as a soft landing?

Consider, for instance, Yuri Gagarin's end-of-voyage dramas in 1961. He had flown quite a lot higher at launch than intended – not to 230km but to 327km, the fault of the second-stage rocket shutting down slightly later than it should have done, forcing the capsule upward. As a consequence, it was clear to Ground Control from very early on in the mission that he was eventually going to land in the wrong place. The only question was, how wrong would that place be?

Inside the capsule, Gagarin himself was kept oblivious to the situation. How could it help him to know? His almost bare cockpit lacked any gauges to tell him how high he had reached. He could only sit and wait, while an old Russian waltz, 'Amur Waves' by Max Kyuss, played into the spacecraft on a loop.*

On re-entry the Vostok's service module and the capsule at first failed to separate, remaining tethered by a solitary bundle of wires. This put the craft into a vicious spin for nearly ten minutes. Finally the rising heat of re-entry burned through the wires, freeing the capsule to descend on its own. Inside, Gagarin could hear crackling and smell burning. Because of the steepness of his re-entry, the g forces were so extreme that he almost blacked out.

'Things began to lose their colour,' he recorded. 'I had to strain to exert myself to focus.'

He kept it together, though. And at 23,000 feet, as per the plan, with a massive bang the hatch behind him was blown off and the cabin abruptly flooded with daylight and freezing cold air. Two seconds later, Gagarin's seat ejected backwards and he was sent spinning into the air.

He was off target by 250km but, by an incredible stroke of luck, he looked down and recognised exactly where he was: there was the Volga river, snaking below, and houses he knew to belong to the village of Saratov.

'Everything here was familiar to me,' he later wrote. 'The fields, the thickets, the roads . . .'

Extraordinarily, Gagarin was descending home through the very skies in which he had originally learned to parachute.

Even now, though, faults and glitches dogged him. His survival

* I thoroughly recommend that you find 'Amur Waves' on Spotify, or similar, and, while it plays, imagine Gagarin circling the globe alone in his tiny spacecraft. I guarantee you an evocative few minutes.

pack – including the inflatable dinghy that might have been useful to him had he landed in the Volga, and the shark repellent that might have been less so – detached from his harness and set off for Earth without him, taking his knife and his first aid kit with it. Then the breathing tube from his helmet got stuck and he spent several minutes wrestling with that as he floated down through the air.

Finally, his back-up parachute, completely unprompted, partially opened. If the back-up had got wrapped around his main parachute, he would have dropped like a stone and known the especially grim irony of being killed by a safety feature.

Fortunately, his functioning parachute was not hindered by his non-functioning one, and Gagarin landed softly in a freshly ploughed field, somewhat alarming a small girl and her grandmother who were out planting potatoes.

'I didn't even feel the landing,' he later recorded. 'I didn't understand at first that I was standing on my legs.'

Gagarin asked the potato-planting grandmother if there was a telephone anywhere nearby. She told him the nearest phone was quite distant, and she suggested he borrow a horse and cart to get there.

For a moment the delicious, era-collapsing possibility arose that the world's first official traveller beyond Earth's bounds would leap from his spacecraft directly into a horse-drawn vehicle. But before that could be arranged, a small convoy of military vehicles from a nearby base, who had seen Gagarin's bright orange parachute in the sky, showed up.

Just as the autograph-hunters had confused him at the launch site, now Gagarin was puzzled as to why everyone in the rescue party was referring to him as Major Gagarin. Who were they talking about? He was First Lieutenant Gagarin.

Not any more he wasn't. First Lieutenant Gagarin had been

promoted while he was aloft, and the news had gone out around the world.

Soon after, Major Gagarin and his wife Valentina were being swept through Moscow in an open-topped car alongside Nikita Khrushchev, with police outriders and thousands of onlookers crammed along the pavements and leaning out of windows to cheer and wave at them.

It was Russia's victory, yet even in that partisan atmosphere the orbiting of the Earth seemed in its magnitude to transcend politics entirely, to be literally and metaphorically above all of that, to be something to catch the whole world in a state of wonder. For all the bitter divisions of the Space Race, this, it seemed, was what space exploration could deliver when it worked: unity.

And then, the following year, 1962, came John Glenn's struggles, oddly parallel in some ways to Gagarin's, give or take a potato field and the threat of a horse-drawn interlude.

As Glenn triumphantly orbited the Earth, what he didn't know was that a sensor reading was deeply troubling Ground Control. Apparently the Mercury's landing bag was no longer locked in position above the heat shield and there was now possibly very little to hold the vital shield onto the craft during re-entry, meaning the shield would likely fall away and Glenn would go into the conflagration of re-entry wildly unprotected.

Like Gagarin, though, Glenn was kept oblivious to his predicament, preoccupied as he was with the closer-to-hand matter of an occasionally malfunctioning flight control system, which would not do what he wanted it to do and was greedily eating into his fuel supply.

Meanwhile, a worried team on the ground at Cape Canaveral racked their brains. Eventually the decision was taken not to jettison the retro-rocket pack after use in the hope that it might help keep the heat shield clamped to the Mercury. The instruction was passed up to Glenn, with no further explanation, and everybody crossed their fingers.

The capsule came hurtling through the upper atmosphere and entered its contactless phase. In a few minutes, the ground controllers would either hear from Glenn or they wouldn't. In the meantime, everybody's only option was to sit there and try not to explode with tension.

The Capcom was Al Shepard, who eventually put out the call.

'Ah, Seven, this is Cape. What's your general condition? Are you feeling pretty well?'

There was a split-second pause that felt like an hour. And then through came the voice of Glenn: 'My condition is good, but that was a real fireball, boy. I had great chunks of that retro pack breaking off all the way through.'

'Very good,' replied Shepard. 'It did break off – is that correct?'

'Roger,' said Glenn.

At the Cape, there was a mass mopping of brows.*

There is a photograph of Glenn taken after his recovery from the Pacific that day, sitting on the deck of USS *Noa* awaiting the helicopter to take him back to land: he's in a NASA flight suit and Ray-Ban Aviators, and his feet are up in a pair of black Converse baseball boots. If you are looking for a definition of 1960s astronaut cool, look no further.

His overt ambition, his raw determination to be the first among equals, and the sense of him as something of an operator earned John Glenn some raised eyebrows from his colleagues along the way. Nevertheless, let's remember that it was Glenn who resisted NASA pressure in favour of protecting his wife, Annie, from the

---

* As it happened, the indicator that lit up to start this panic had been triggered by nothing more than a loose switch. Glenn's landing bag had been fine along.

media storm that Vice-President Lyndon Johnson was set to bring down on her by visiting her at home in the wake of one of Glenn's multiple scrubbed launches. Annie sometimes stuttered when speaking publicly and didn't want this exposure forced on her. Glenn told her, over the phone from the launch site, that it was her home and that if she felt she needed to, she should send the Vice-President away – which Annie duly did, much to Johnson's fury and NASA's discomfort. And let's remember that it was Glenn who insisted on all the other Mercury Seven astronauts being part of his motorcade and ticker-tape parade through New York following that glorious, seat-of-the-pants return to Earth. Ambitious Glenn might have been, but he recognised that the true glory was never any single astronaut's alone.

Sometimes a spacecraft returns in triumph and sometimes a spacecraft returns in a mixture of triumph and slapstick, which is what happened when the crew of Apollo 12 witnessed a solar eclipse on their way home and recorded it on camera, only for that same camera to fly across the capsule with the impact of splashdown and clonk Alan Bean on the head, leaving him with six stitches in a 2cm cut above his right eyebrow.*

And sometimes the narrative of a spacecraft's return is richer than you would wish it. When Yuri Malenchenko returned from the ISS with Peggy Whitson and Yi So-yeon, a South Korean biochemist, on TMA-11 in 2008, the service module at the rear of the Soyuz failed to separate at first, putting the craft at a dangerously wrong angle for re-entry. The segment did eventually

---

* This head injury merely continued Bean's run of bad luck with cameras. During the Apollo 12 mission he had set up a colour television camera on the Moon, with the US networks all greedily poised for colour coverage, only to point it directly at the Sun and destroy its vital tubes. 'Here's the TV,' Bean had laconically remarked as he completed the installation. 'And it's pointing towards the Sun. That's bad.'

break away and the craft righted itself, but its hatch and antenna were badly scalded and the crew had noticed something no astronaut wants to see during a landing – smoke in the capsule.

Meanwhile, the separation episode had set them on a ballistic re-entry path of 30 degrees and led the craft to experience 8 gs, twice the force they would have been expecting to endure. They were plunging and then parachuting through the air for twenty-three minutes and they landed 300 miles short of the intended target zone.*

'I saw 8.2 gs on the meter and it was pretty dramatic,' Whitson said afterwards. She had been on board the ISS for six months. 'Gravity is not my friend right now,' she commented right after the flight. 'And 8 gs is *definitely* not my friend.'

But of course, it's the destiny of all astronauts, however they return, to come down to Earth with a bump. The fact is, what we call 'a landing' is always a crash – and that's equally true whether we're talking about a spacecraft falling to the ground from the atmosphere, or a commercial plane dropping onto a runway, or you or I jumping down off a chair. It's a basic law of physics and gravity that the only way anything can return to the ground from the air is by crashing onto it. The question is only ever how much degree of control you can exercise over the crash.

And sometimes that control is as tight as can be, and a comfortable experience ensues.

And sometimes there is no control at all.

---

* Entirely embarrassing himself, the Russian Space Agency chief Anatoly Perminov described the accident in a press conference as bearing out an old naval superstition about women (Whitson and Yi So-yeon) outnumbering men (Malenchenko) on a ship and vowed to take measures to prevent that happening again. I repeat that this was in 2008, not, as you might have assumed, 1608.

III. DEVIL MACHINES

People would say of Vladimir Komarov that he was born looking up. Growing up in Moscow in the 1930s, he obsessively watched planes from the window of his family's apartment and learned to distinguish them by sound alone. And he made his own models of them, cutting propellers out of tin can lids. Nobody was surprised when he ended up in the Air Force and qualified as a flight test engineer, nor really when he was summoned to join the first Soviet cosmonaut draft.

At that point Komarov moved to Star City with his wife, Valentina, and their son and daughter. He was thirty-two, the second oldest in that twenty-strong cosmonaut pack, and soon regarded by his peers as an elder statesman. He helped the younger ones with the studies that were part of their training – he and Pavel Belyayev were referred to as 'the Professors' – and he developed a particularly close friendship with Yuri Gagarin. A photograph shows the pair of them out hunting together – berets on, double-barrelled shotguns in hand, broad smiles on their faces.

Having successfully commanded a mission in tight conditions on Voskhod 1 in 1964 – that bravado launch of a three-member crew, mentioned earlier – Komarov was appointed, three years later, to fly on Soyuz 1.

Soyuz in English is 'union' or 'alliance' – a reference, obviously, to the Soviet Union itself, but also a pun for the purposes of this particular mission. The plan was to get Komarov in orbit, then to launch a second Soyuz, containing two further cosmonauts. The two craft would rendezvous in space and dock, then Komarov would crawl into the second capsule, and one of the other cosmonauts would crawl the other way. And then, with pilots swapped, the capsules would separate and fly home.

A union in orbit, plus a cosmonaut seat-switch? Here, clearly, was a mission with some high showbusiness value, not unintentionally planned to coincide with the fiftieth anniversary of the Russian Revolution, but also designed to tweak NASA's nerves by conveying the absolute capability of the Soyuz to carry Soviet cosmonauts to the Moon sometime very soon.

And, for maximum impact, no news of this space spectacular had been released in advance. The West could read about it in the papers afterwards – read about it and weep.

For all these bravura plans, though, behind the scenes Soyuz 1 was beset by problems. Four preliminary launches had revealed a mass of bugs and glitches. One investigation of the spacecraft appears to have identified as many as 203 problems that needed addressing. Komarov, according to some sources, described the Soyuz as a 'Devil machine! Nothing I lay my hands on works.' Gagarin, Komarov's back-up on this upcoming flight, is alleged to have written to his seniors at one point, on behalf of all the cosmonauts, expressing his anxieties about the craft and suggesting the mission be delayed.

But a major national anniversary is a major national anniversary. Gagarin's qualms were ignored, and the launch remained set for 23 April 1967.

Not long after lift-off, Komarov reported to Ground Control that the left solar panel on the spacecraft had not deployed. Nothing he could do, including drumming his feet against the inside wall of the capsule, would free it. He was now inside an asymmetrical craft whose power supply had halved. Moreover, the undeployed panel was interfering with the sensors required for the Soyuz's attitude control. Komarov was having to try to keep the craft stable manually with bursts of propellant from the thrusters, but that was consuming his fuel at an unsustainable rate.

Ground Control realised that they were already looking at a potential disaster. After five orbits, they cancelled the launch of the rendezvous and docking vehicle and turned their efforts to the single task of getting the Soyuz, and Komarov, home.

Automatic re-entry was set for the eighteenth orbit but the combination of the craft's asymmetry and its faulty attitude control pulled the Soyuz off its path and, in response, the retro engines shut down at the critical point. Komarov was given instructions to go round again and perform a manual re-entry on the nineteenth orbit. Somehow, in the face of drastically diminished odds, he managed it, holding his nerve and doing what the automated system couldn't to bring the underpowered and lopsided spacecraft back through the atmosphere at a survivable angle and out the other side. It was an extraordinary achievement.

All he had to do now was wait for the parachutes to deploy and float him back to the surface.

Except that the parachutes didn't work. The main parachute failed to deploy. The secondary chute did pop out but then got tangled around the failed main parachute and didn't open. Soyuz 1 hit the ground on farmland in Orsk at huge speed and exploded.

Sometimes no amount of bravery and ingenuity will win the day against dumb luck. Just three months after the Apollo 1 crew lost their lives on the ground, spaceflight had claimed its first life from the air.

On 25 April, Komarov's cosmonaut colleagues published a tribute to him. It read:

For the pioneers it is always more difficult. They tread the unknown paths and these paths are not straight, they have sharp turns, surprises and dangers. But anyone who takes the pathway into orbit never wants to leave it. And no matter what difficulties or obstacles there are, they are never strong enough to deflect such a man from his chosen path. While his heart beats in his chest, a cosmonaut will always continue to challenge the universe. Vladimir Komarov was one of the first on this treacherous path.

The first and, of course, not the last.

On 30 June 1971, the crew of Soyuz 11, Georgy Dobrovolsky, Vladislav Volkov and Viktor Patsayev, were returning from the Salyut 1 space station in triumph. Two years previously, Armstrong had descended to the Moon and America had definitively clinched the Space Race, devastating the Soviet space agency, who could then only look on as a succession of American crews proudly headed the same way. So many times during that race Soviet cosmonauts had streaked ahead of their American counterparts, only to be financially and technically outmuscled in the sprint to the big prize. The construction of weapons in order to keep pace with America in that other ongoing event, the Nuclear Arms Race, had inevitably leached funds from the space programme and caused the Soviet Union to drop back as the sixties wore on.

The final blow to Soviet hopes had come on 3 July 1969 when, in an uncrewed test, a prototype N-1 rocket, in which all the agency's hopes of carrying a Moon-bound payload into orbit had been invested, exploded five seconds after launch at Baikonur, destroying the pad, flinging shards 10km and causing windows to shatter in buildings 40km away. Nikolai Kamanin, the cosmonauts' head of training, recorded in his diary: 'The failure has put us back another one or one and a half years.' Less than a fortnight later, America was on the Moon.

Here, though, was a chance to regroup and restore some pride – in the new category of crewed orbital space stations. Dobrovolsky, Volkov and Patsayev had just become the first team to work aboard an orbital laboratory. They had spent over twenty-three days in space – a duration record that America would not match for another two years. They had appeared nightly on Soviet television screens, Volkov in particular becoming something of a star and national heart-throb. And now they were back on Earth, where accolades awaited them after, fittingly, an on-time and on-target landing.

The capsule reached Kazakhstan at 2.16 a.m. Moscow time. Nikolai Kamanin, the head of the cosmonaut corps, waited expectantly for a phone call bringing news from the site.

There was a system of numbers in operation for immediate health updates on returning space crews: '5' meant the cosmonaut was in excellent health; '4' meant that they were in good condition; '3' meant that they had suffered injuries; '2' meant that the injuries were severe.

Kamanin listened in horror as he heard the recitation from the staff on the ground.

'One, one, one.'

The rescuers had found the cosmonauts in a seemingly entirely intact capsule, still in their seats, with bruised-looking faces and blood around their noses and ears. It was estimated that they had been dead for at least an hour. A valve in the craft had come open during the orbital descent phase, soon after separation. The rapid decompression had drained the capsule of its air supply in under a minute and sent bubbles of nitrogen through the cosmonauts' blood. Had they been issued with spacesuits, they would almost certainly have survived.

Their three bodies were laid in state in the Central House of the Soviet Army in Moscow, in open coffins on a giant catafalque swathed in flowers. Tens of thousands of stunned Muscovites filed past silently in despair, as did Leonid Brezhnev, the leader of the Soviet Union, who was seen to hide his face behind his hand at one point.

In more than sixty years of spaceflight, and after more than 370 crewed flights into the ruthlessly hostile environment of space, Dobrovolsky, Volkov and Patsayev remain the only humans to have died above the Earth's atmosphere.

Long may they hold that distinction.

IV. ALL THE WAY DOWN

My Commander on the ISS, Scott Kelly, described re-entry on the way back down to Earth as 'like going over Niagara Falls in a barrel – that's on fire'.

And he wasn't exaggerating.

Here's how that might go for you, assuming you were departing from the ISS and heading home in a Soyuz capsule.

Pushing back gently from the space station in the capsule, your initial feeling is going to be one of immense vulnerability. It's the huge contrast, obviously, between the scale of the craft you have just left and the craft you are now tightly belted into with your two colleagues. The station suddenly seems like a multi-deck cruise ship in contrast to the rubber dinghy of the capsule.

But somehow it's also much more than that – a realisation, as you move away from it, about what the station has come to mean to you during the time you've spent on board: the protection it has afforded you and the security you have drawn from it. It's been your place of refuge for maybe six months or more, your safe place, and, in a way that you might not have predicted, it's actually a wrench to leave.

Probably the next thing you'll feel, as the thrusters propel you beyond the ISS's golden solar arrays and out into space, is an overwhelming tiredness. The chances are you've been awake for hours. Getting everything set for departure is complicated and sapping. There will have been a crew sleep written into the schedule somewhere relatively recently, but the chances that you will have managed to use any of it for actual sleeping are remote, given the anticipation that will have been surging through your nerves at that point.

It's like the phenomenon of the launch-pad doze that we witnessed earlier. You seem to have been rushing around strenuously for hours, and now here you are in a small, warm, quiet capsule, softly floating through space to the gentle soundtrack of the humming fans. The urge to allow your eyes to close and just drift off to sleep for a while is powerful.

You'll be keenly awake soon enough, though – probably on the second orbit. This is the point at which you're going to find out whether your tiny spacecraft's main engine still works. The capsule has been sitting in space for quite a while – parked on the ISS's drive, as it were. It's going to be nice to find out that its engine still turns over when asked to.

Things you'll be hoping to notice here: a low growling sound and a new sense that the capsule is being pushed along firmly from behind. If those things are present and correct, it means you have a main engine to work with and you can breathe again.

This engine burn will need to be precisely timed. If it goes on even slightly too long, you will be too steep when you re-enter, making things very uncomfortable indeed (witness the experience of Yuri Malenchenko, Yi So-yeon and Peggy Whitson, above). If it doesn't go on for long enough, however, you'll miss your entry and find yourself skimming through the upper layers of the atmosphere like a stone off a pond and heading out into space again.

Four minutes and thirty-seven seconds should do it. No more, no less. That will have you descending on the right trajectory with another half of the planet still to cover before your re-entry point.

Ahead of that, though, a further quarter of the way around the planet and descending, it will be time for the spacecraft to separate, breaking into three. The habitation module is sent off in one direction and the service module is sent off in another, while the part you're sitting in, the descent module, continues its journey.

You'll need to be ready for some serious noise at this point because those sections don't simply uncouple; they have to be blown apart. So get set to feel as though the capsule is abruptly coming under machine-gun fire – fourteen pyrotechnic bolts popping off in rapid sequence.

And those explosions are taking place through the wall just behind your ear, each one of them giving the capsule a shake. There'll be no sleeping through that.

All this time you've been descending. Earth is only about 120 miles below you now, and the capsule is in a gentle tumble, end over end, performing a lazy somersault through space. You can see parts of the world slide slowly by the window, alternating with passages of darkness as the capsule turns.

Those somersaults will bring you all the way down into the upper layers of the Earth's atmosphere where, praise be to natural aerodynamics, the craft will cease to tumble and orient itself heat shield forward. Or you very much hope so, anyway.

Here's where you will notice the first mild onset of g forces as your long-lost friend gravity quietly begins to reintroduce itself to you. Time to stow your checklists in the pocket provided, push yourself firmly back into your seat, and pull your harnesses around you as tight as they will go. And please be informed that the bathrooms are now out of use.*

---

* This is a joke. There are no bathrooms. There is, however, a Maximum Absorbency Garment (MAG) worn by all astronauts which is available for use should you so desire.

Now it's time for sparks to fly – just one or two at first, flicking past the window beside you, then a steadily growing number until they are beyond counting. Nothing to concern you: those are just bits of burning spacecraft. You can't expect to pass through temperatures as high as 1,700°C without sacrificing some parts of the vehicle you're travelling in. Nor without the cabin heating up, which is definitely happening right now because you can feel yourself starting to sweat.

The window beside you is beginning to scorch, charring and browning over, and there will be no more seeing anything out of it, even assuming you could put your head up to it to look. However the g forces have been slowly building, through 2 and 3 to 4, 5 . . . and the sound of the air rushing past has grown from a whisper to a whistle to a roar akin to a jet's engine, and now you are pinned back in your seat and extremely aware of the fact that you are travelling several times faster than the speed of sound.

You will know you are out the other side and officially back in Earth's atmosphere when those g forces diminish slightly. And now it's time to prepare yourself for the popping of the braking chutes. Two pilot chutes are going to drag out the drogue chute to retard your flight, and the consequence of those chutes coming out is going to be harrowingly violent. The capsule is going to start flinging around wildly on the end of its new strings, even as it continues to plummet, bobbling and throwing itself in circles, and this is going to go on for about twenty seconds.

And I don't think it's any exaggeration to say that, if you hadn't been trained to expect it, you would be assuming during those twenty seconds that something was terribly wrong and that your life was probably about to come to an end.

Be assured, though, that the wild agitation is going to cease eventually, and that's the signal that at any moment now the main parachute is going to pop out. Sometimes the emergence of that parachute hands the capsule a mighty jolt too, but other times you'll get lucky and barely feel a thing, which is what happened to me.

And now you can prepare yourself for about a quarter of an hour of relatively peaceful downward drifting, give or take the sweat that's stinging your eyes and which you can't clear because your visor is down. But you are safe in the knowledge that your antenna is now letting the search and rescue teams know that you're on the way.

Of course, you've still got the actual landing to get through. And that, you've been warned, is going to be uniquely rough.

Brace yourself, then: arms across your chest, neck back, head steady, tongue well back from your teeth.

There will be two sharp bangs as the soft landing thrusters fire, about half a metre above the ground, and those bangs are your cue to ready yourself for landing – which is to say (for the scientific reasons mentioned above) for the controlled crash.

I can only describe it as like getting cracked with a baseball bat – an almighty body blow as the capsule slams into the surface. Then it bounces and rolls over, with you rolling inside it, and skids along the ground until it eventually gets dragged to a halt by the parachute.

And then all is still and it's over. Leaving you in whatever attitude the capsule has come to rest – in my case, with the capsule on its side and with me lying at the top, reaching an arm above me to stop the maps and checklists in the storage pocket from dropping out onto my colleagues below.

Drenched in sweat, you will also now be fully aware of gravity once again featuring in your life. Just the slightest move of your head to one side will cause the world to spin. You will want to concentrate on looking directly ahead of you for a little while until your head remembers how this all works.

Soon, though, the search and rescue team will be outside, righting the capsule and opening the hatch. Now, a lot of people will tell you about the wonders of returning to Earth after a period away in the sensory deprivation of space, and especially about the olfactory wonders. And, of course, they're right. But for me, those much-missed smells – fresh air, rain, pizza – would provide a source of delight later. From the instant aftermath of landing, my overwhelming memories are of the smell of burned grass mingled with the even more bitter smell of burned spacecraft and the actually rather nasty smell of overheated electrics.

Plus, as a large figure reached into the capsule to pull me out, a top-note of burly crew guy.

They had been in position and waiting for us out on the Kazakh steppe in the full heat of mid-summer for a few hours, those search and rescue guys. It had been a long day for them too, clearly. A long, sweaty day.

V. 'COLUMBIA, HOUSTON: COMM CHECK'

The spacecraft that seemed to have got the whole landing business well and truly sorted out was the Space Shuttle. Not for those Shuttle crews a bruising bounce across the Kazakh steppe, nor even a bracing slap against the waves of the Pacific. They would touch the runway on tyres at a relatively civilised 200mph, oriented like any commercial airliner, and then brake to a halt.

Yet, unlike any plane you are likely to catch, the Shuttle would, not long before that moment, have been found 10 miles above the ground, arriving nose-down and going sub-sonic – those two booms heard the length of the Florida coast – and would then have adopted a 20-degree glide slope at 300mph.

And yes, John Young likened it to flying 'a lead safe with its door open', the open door providing you with what you had in the way of lift. And yet the Shuttle commanders would also tell you that, in that glide phase, it flew as responsively as a fighter jet – albeit a 100-ton fighter jet.

Anyway, at 1,500 feet above the ground, the Commander would start pulling the nose up and the Shuttle would decelerate. At 300 feet, the landing gear would go down, and soon after that the craft would drop onto the runway, scampering along with its rudder speed brake on until the drag parachute popped out and the wheels came to a complete stop.

And at that point everyone would stand up, get their bags from the overhead lockers and troop off.

Or something close to that.

Certainly Shuttle landing days at the Kennedy Space Center had different atmospheres from Shuttle launch days. Arrivals generally invite lighter emotions than departures in any case, but memories of the *Challenger* explosion in 1986 were bound to tint every Shuttle launch thereafter. Landing days could float relatively free of that association and therefore tended to be more relaxed – a party blooming amid the general good faith in the extraordinarily trustworthy routine-ness of it all.

And, in as much as the arrival of any craft from space can be, Shuttle landings really *did* get to feel trustworthy and routine. By February 2003, the Shuttle had come down 111 times and, with the exception of one blown tyre on STS-51-D (the fourth flight of *Discovery* in 1985) and one thankfully harmless touchdown 600 feet short of the runway in high winds on the dry lake bed at Edwards Air Force Base, landing had been blissfully free of problems.

But of course, even the astonishingly cunning Shuttle couldn't bypass the trauma of the re-entry phase. The plummet through the atmosphere was, by all accounts, quite something to witness through the windows of the Shuttle's flight deck, which offered the viewer a rare chance to position themselves at the centre of a fireball for a few minutes.

Astronauts have a term for these more extreme experiences: 'interesting'. 'That's where it gets interesting,' they'll say, selecting that word over others which people who aren't astronauts might reasonably choose, such as 'intense', or 'violent', or 'utterly terrifying'.

The Shuttle would be travelling at over 17,000mph until the Commander lowered its speed by just 200mph, which was enough to cause it to fall out of orbit. Then there would be a retro burn over Australia with a view to dropping into the sky above Florida about an hour later.

That was the last use the Shuttle would make of power: it was all freefall and gliding from there, the craft now slicing through the atmosphere at 1,500°C. During its gliding re-entry, the Shuttle would pull a relatively sedate 1.7 g – a breeze compared with the 'falling brick' capsule design and punishing ballistic re-entry. From a seat on the deck you would apparently see the cabin flicker and flash, as if the paparazzi were outside. And then you would see the envelope of plasma glow a salmon pink colour, and wrap round the craft like tentacles, toning down to a dull white as the atmosphere got thicker and the craft slowed, and then eventually releasing it altogether.

However, by 2003 there had been one serious incident involving the Space Shuttle in this phase of the return journey. During the launch of flight STS-27 in 1988, *Atlantis* had been struck by insulating material flying off one of the solid rocket boosters and had sustained damage to some of the silica tiles affixed to its aluminium frame to insulate it and deflect heat.*

Mike Mullane was the crew member charged with using the Shuttle's robotic arm as the craft flew in space to run a camera over the damage, and both he and the flight's Commander, Hoot Gibson, were deeply shocked by the quantity of damage they saw, and seriously worried about the consequences for *Atlantis* and themselves in the heat of re-entry.

Mission Control, however, examining the same footage, repeatedly insisted that there was no problem.

Was that genuinely the case, or did Mission Control essentially mean that there was no problem *that the crew could do anything about*? Mullane, Gibson and their colleagues simply had to wait it out, complete their mission, and try to relax about it. As Gibson memorably put it: 'No reason to die all tensed up.'

After a harrowing re-entry, which Mullane spent on the floor of the flight deck rather than in his assigned seat on the enclosed mid-deck, so that, as he put it, 'at least I could die looking out a window', *Atlantis* landed normally. On the ground, engineers discovered 700 compromised tiles in an area that ran half of the Shuttle's length – an area that just happened to be over an antenna which had clearly provided extra thickness and protection to the vulnerable aluminium frame.

Atlantis accordingly won the distinction of becoming the most damage-riddled spacecraft ever to make it home, and NASA breathed again. As Mullane put it, 'We had taken 700 bullets and lived to talk about it.'

The crew of *Columbia* would not be so lucky.

Mission STS-107 was commanded by Rick Husband, an Air Force test pilot and mechanical engineering graduate who had

---

* Those silica tiles earned the Shuttle the nickname the 'Glass Rocket'.

already served a mission on the ISS. Its pilot was William McCool, who would send the following message to Earth during the sixteen-day flight:

> From our orbital vantage point, we observe an Earth without borders, full of peace, beauty and magnificence, and we pray that humanity as a whole can imagine a borderless world as we see it and strive to live as one in peace.

There were also four Mission Specialists on board: Michael Anderson, who had flown Space Shuttle *Endeavour* to the Mir space station in 1998 and was the ninth African American to go to space; Kalpana Chawla, known as KC, who was born in India and educated in Texas and Colorado and who, on *Columbia* in 1997, had become the first Indian-born woman in space; Laurel Clark, a medical doctor and former naval submarine officer from Iowa, on her first spaceflight; and David Brown from Arlington, Virginia, another rookie and a former naval flight surgeon. And there was a Payload Specialist, Colonel Ilan Ramon, an Israeli Air Force fighter pilot, on board to conduct an experiment observing dust storms in the Mediterranean region. He was the first Israeli astronaut.

This was probably as talented and as diverse a Space Shuttle crew as NASA had ever selected. And none of them made it back.

In the memories of people who were at Kennedy and Houston on that Saturday morning in February 2003, the non-return of STS-107 remains a matter of ominous absences. It's about what people didn't hear, what people didn't see, what didn't happen. It's about the sonic double-boom that never came. It's about sensor read-outs suddenly dropping out on screens in Mission Control in Houston. It's about the lack of anything on the tracking radar at Kennedy. It's about the awful silence that greeted the repeated calls of Charles Hobaugh, the Capcom.

'*Columbia*, Houston: comm check.'

Nothing.

'*Columbia*, Houston: comm check.'

Nothing.

'*Columbia*, Houston: comm check.'

And it's about the landing strip, still eerily quiet at *Columbia*'s scheduled time of arrival – 9.16 a.m. Eastern Time.

Elsewhere, witnesses have more visceral recollections. *Columbia* came apart while travelling at 11,000mph, 181,000 feet above the Texan cities of Corsicana and Palestine, to the southeast of Dallas. People below the flight path heard booms and cracklings, rumbles and bangs, variously interpreted as a gas pipe explosion, a train crash and an act of terrorism. Some saw smoke trails in the sky – some of them straight, some spiralling, seemingly going in all directions.

Then the debris began to fall, pattering into a reservoir, thudding into the ground. Wreckage fell from the skies over Texas and Louisiana for thirty minutes on a path that ran for 250 miles.

As with *Atlantis* on STS-27, the damage had been sustained during the launch phase. A lump of foam, about the size of a laptop computer, had broken off the external tank a minute and a half into the flight, striking the left wing of the Shuttle and shattering into tiny pieces. That blow had in no way compromised the operation of the spacecraft throughout its entirely successful sixteen-day mission, but it had left an undetected hole in the covering of the wing – a fatal flaw which *Columbia* carried obliviously throughout its fortnight in space and which then, during the re-entry phase, allowed superheated gases to breach the craft and ultimately overwhelm it.

And once again, seventeen years after *Challenger*, NASA

found itself picking up the pieces – incredibly, 83,000 individual items of debris were eventually recovered, just under 40 per cent of the Shuttle – in an effort to understand what had happened and intent on using that understanding to go forward.

But this time there could be no going forward. Compounding the *Challenger* disaster, the disintegration of *Columbia* made the gradual retirement of the orbital lander inevitable. On 21 July 2011, *Atlantis* glided onto the runway at Kennedy and reached wheel-stop for the final time, and the era of gentle returns came to an end.

VI. DON'T ASK ME, BABE

The crew of Apollo 11 splashed down in the Pacific Ocean early in the morning of 24 July 1969, 200 miles off course, and for a very good reason. A US weather satellite had picked up brewing storms in the original landing zone, most likely bringing thunder and winds high enough to tear the capsule's parachutes apart. Which, with the world awaiting a heroes' return, would have produced an anti-climax to say the least. The re-entry path and search and rescue plans were quickly adjusted, and the recovery ship USS *Hornet* was only 13 miles away, poised to move in, when the capsule dropped into the water.

The crew still had to don biological isolation outfits and head off to quarantine for three weeks. But after that, the celebrations could properly begin.

Flanked by police outriders, Neil Armstrong, Buzz Aldrin and Michael Collins rode in open-top cars through clouds of ticker-tape in New York, Chicago and, finally, Los Angeles where President Nixon, who had spoken to them on the Moon – 'the most historic telephone call ever made from the White House' he had called it – now staged a state banquet for them, enabling them to mingle with another superstar of aviation (Charles Lindbergh)* and a glittering selection of Hollywood royalty (Fred Astaire, Joan Crawford, Bob Hope).

Then commenced the 'Giant Leap' tour. Think Taylor Swift and then some: twenty-nine cities in twenty-four countries in thirty-eight days, using Air Force 2, Vice-President Spiro Agnew's plane, and kicking off in Mexico City, where hundreds of thousands thronged the streets. And then it was on to Bogotá, Brasilia, Buenos Aires, Rio de Janeiro, Las Palmas, Madrid, Paris . . . By the time they had finished with Perth and Sydney and flown to Tokyo, via Seoul and Agana, they had crossed the equator six times, given twenty-two press conferences and shaken hands with dignitaries on every continent, including the King and Queen of Belgium, the Shah of Iran, the Emperor of Japan, Queen Elizabeth II and the Pope. In Dhaka, Pakistan – now the capital of Bangladesh – a million people turned out; in Mumbai (then Bombay) in India, 1.5 million.

In each city a briefing sheet had ensured that these travelling Americans were mindful of local customs and in tune with the protocol, but by far the best was the parody document prepared for the tour's final, homecoming stop-over in Washington DC:

1. The water is drinkable, though it is not the most popular native drink.
2. You can always expect student demonstrations.
3. Never turn your back on the President.
4. Never be seen with the Vice-President.

* In 1927, Lindbergh flew the single-seater *Spirit of St Louis* from New York to Paris, the first non-stop flight between those cities. It took him 33.5 hours. When he met the Apollo 11 crew in the summer of 1969, Lindbergh was sixty-seven.

5. If you leave your shoes outside the door, they will be stolen.

6. It is unsafe to walk on the street after dark.

'So, what was it really like?' This was the question they heard again and again – the question, indeed, that they would be hearing for the rest of their lives. They had been to the Moon, and people's curiosity about it was apparently inexhaustible. The public appeared amply to share Walter Cronkite's feeling, voiced in the quotation at the beginning of this chapter, that these returning voyagers from hitherto untouched realms must be in possession of *secrets*; 'secrets that we will never entirely know', Cronkite said, though that wasn't going to deter people from insisting that they be told them.

All the Moon crews would know the strain of this, in varying degrees. Having had the experience, the pressure was now on to account for it – to put it into words. Did this burden straightforwardly play to these astronauts' strengths? It was an open question.

The Apollo astronauts were test pilots, predominantly, and test pilots have a professional duty to keep the tiny details in view and find things to be unimpressed by. The sweeping metaphysical view with fully fledged epic overtones . . . well, habitually, maybe that tended to be less their thing.

Certainly, during the Moon missions themselves, attempts to have philosophical discussions about *what it all might ultimately mean* rarely seem to have taken off. Perhaps there wasn't much time for that. But perhaps there wasn't too much inclination, either.

During Apollo 10, for example, on his first lunar trip, Gene Cernan found himself with a rare free moment for reflection as he gazed out of the window at the Moon, which they were currently orbiting. He turned to John Young beside him and Ground Control then heard the two of them have the following exchange:

Cernan: 'Hey, let me ask you a question. Where do you suppose a planet like this comes from? Do you suppose it broke away from the – away from the Earth, like a lot of people say?'

Young: 'Don't ask me, babe.'

Cernan: 'It sure looks different.'

Young: 'I ain't no cosmologist. I don't care about nothing like that.'

Cernan: 'Sure looks different.'

And of all these astronauts, perhaps none was less inclined to flights of philosophical fancy than Neil Armstrong – although, ironically, it was Armstrong who had composed arguably the most resonantly philosophical thought on the lunar surface when he mused that, from his privileged vantage point, he could hold up his thumb and entirely obscure behind it the world and everything he knew.*

Still, you had to admire the standard response that Pete Conrad eventually developed for the 'What was it really like?' question: 'Super! Really enjoyed it!' And you could also relate to the frustration voiced by Michael Collins: 'If one more fat cigar smoker blows smoke in my face and yells at me, "What was it really like up there?", I think I may bury my fist in his flabby gut. I have had it with the same question over and over again.'

It all drove Buzz Aldrin over the edge for a while. Just over a year after walking on the Moon, Aldrin suffered what he himself described as 'a good old American nervous breakdown', and, in what was at the time a closely guarded secret, he underwent a

---

* 'It suddenly struck me,' Armstrong later said, 'that that tiny pea, pretty and blue, was the Earth. I put up my thumb and shut one eye, and my thumb blotted out the planet Earth. I didn't feel like a giant. I felt very, very small.'

month of psychiatric treatment at an Air Force hospital in San Antonio. And at least part of Aldrin's distress seems to have been the demand that he encapsulate his experience in words for people, along with a constant sense that he was failing to do so. He was found at one point weeping in a corridor behind a lecture hall.

But there were happier outcomes, and those who coped more pragmatically with the aftermath of their Apollo experience, and the heat of the fame that it thrust upon them. It was Alan Bean's plausible contention that 'everyone who went to the Moon came back more like they already were'. On his return from Apollo 12, Bean spent a day in a shopping mall, eating ice cream and watching people go by, in what seemed to him a renewed state of wonder for the world and all it offered. He stated his opinion that there was nothing that could be seen through a telescope that was as beautiful as the Earth. 'We've been given paradise to live in,' he said. 'I think about it every day.'

And then he took up painting, and painted the Moon, and only the Moon, again and again, mixing Moon dust into his pigments.

Also in Bean's category of 'like themselves, only more so' was Alan Shepard. On Apollo 14, Shepard leaned back to look at the Earth and found himself crying. He was the only one of the twelve Moonwalkers to admit to shedding tears on the Moon. When he came back, Shepard said, 'I suppose I'm a little nicer than I used to be.'

But then there was Ed Mitchell, arguably of all the Moonwalkers the one most overtly changed by the experience. As an astronaut on Apollo 14, Mitchell said, he had been 'a bear going over the mountain to see what he could see'. Now he spoke of how the sight of the whole Earth gave him 'an instant global consciousness . . . an intense dissatisfaction with the state of the world and a compulsion to do something about it. From out there on the Moon, international politics look so petty. You want to grab a politician by the scruff of the neck and drag him a quarter of a million miles out and say, "Look at that, you son of a bitch."'

Mitchell now saw himself as a piece of the universe, at one with the stars and planets – possibly even at one with the son-of-a-bitch politicians. He felt, he said, a 'desire to live life to the fullest, to acquire more knowledge, to abandon the economic treadmill'. He sought spiritual connection via Buddhist, Hindu and Native American writings.

He also declared a belief in UFOs, claimed the government had covered up alien visits, and maintained that he had been cured of renal cancer by a teenage remote healer from Vancouver named Adam Dreamhealer.

Mitchell died in a hospice in West Palm Beach in 2016 at the age of eighty-five.

Rusty Schweickart, too, seemed to find a new sense of universality out there in space. On Apollo 9, it was Schweickart's duty to perform a stand-up EVA – the first EVA of the Apollo programme – on the so-called 'porch' of the Lunar Module and test the life support systems that the astronauts would eventually wear on the Moon. Talking about that experience in an interview in 1972, Schweickart said, 'I completely lost my identity as an American astronaut. I felt a part of everyone and everything sweeping past me below.' Boundaries became irrelevant to him, he said – not just national boundaries, but the boundaries between himself and other people, between himself and other things. He felt now that when he spoke of himself having experiences, what he really meant was 'life' having those experiences.

Schweickart grew his red hair a little longer than an astronaut would conventionally wear it, added mutton-chop sideburns, became interested in Zen Buddhism, and gave a conducted tour of the Manned Space Center to the Maharishi Mahesh Yogi, with whom the Beatles had studied transcendental meditation, and whom Schweickart had met at a symposium at the Massachusetts Institute of Technology. Later, allegedly, the Maharishi adjourned to Schweickart's home and the pair of them walked barefoot together on the lawn.

And then there was Jim Irwin, who walked on the Moon with Apollo 15 and, along with Dave Scott, discovered the Genesis Rock – or '15415' to give it NASA's less poetic label. It was 4 billion years old – as old as the solar system – and it made a vast difference to our understanding of how and when the Moon and Earth formed. The Genesis Rock was the single most important geological discovery of the Apollo missions.

Working on the lunar surface with Scott, Irwin only lightly referred to a sense he had of God's presence.

Scott had just said to him, 'Look at the mountains today, Jim, when they're all sunlit. Isn't that beautiful?'

Irwin replied, 'Dave, I'm reminded of a favourite Biblical passage from Psalms. I look unto the hills from whence cometh my help. But, of course, we get quite a bit from Houston, too.'

The two of them returned to their task without further comment.

But in fact Irwin later said that he had been powerfully aware, as never before, of God during those EVAs on the Moon – to the point where he had looked over his shoulder at one point and expected to see God standing there. Irwin started a mission on his return, a Baptist ministry called High Flight whose ministers billed themselves as 'goodwill ambassadors for the Prince of Peace'. In 1973 Irwin's faith seems to have led him to Mount Ararat in eastern Turkey to search for the remains of Noah's Ark.*

---

* According to the account of the Noah story in the Book of Genesis, the ark eventually came to rest 'on the mountains of Armenia', now Turkey. The

As these various reactions to the Moon experience emerged, something inevitably shifted in our perception of astronauts – of who they were and what they were. They had been presented to the world as superheroes on the one hand, and on the other hand as largely emotionless, mission-focused automata. Neither of these images had bestowed upon them much in the way of humanity. But a greater truth had eventually emerged – a more mixed picture, a more human picture. As an article by Howard Muson in the *New York Times* in 1972 put it:

> They turned out to be more interesting than we thought. To our relief – we might have predicted – they now seem much like us. They cheat a little when presented with too great temptation (witness the brisk trade in postal covers that were taken to the Moon and other articles that were autographed by astronauts);* they get divorces from their wives† . . . they see a psychiatrist when necessary . . . they get interested in freaky things like ESP, Eastern religions, poetry, one-world-ism and brotherhood . . . and like everyone else

---

ancient Jewish text the Book of Jubilees more specifically locates that resting place on the peak of Lubar, a mountain of Ararat. Irwin's search for the wreck was unsuccessful, as many searches had been before him.

* The crew of Apollo 15 in 1971 were the guilty parties here, having struck a pre-flight deal worth $7,500 each with a German stamp dealer in exchange for carting commemorative envelopes to the Moon and back. NASA were not at all amused by this attempted cash-in, and the tightest reins on what travels to space in astronauts' personal luggage have been in place ever since.

† They certainly did. Of the thirty astronauts who were the Mercury Seven, the New Nine and the Fourteen (the 1963 intake), only seven had marriages that survived. Perhaps in the history of humankind, only *Strictly Come Dancing* has posed such a grave threat to the bonds of marriage as becoming an astronaut in the sixties did.

they complain about bureaucracy . . . in language surprisingly free of gobbledygook.

All of the Apollo astronauts had to find a way to live with the monumental nature of what they had achieved, to cope with the fame and exposure they received as the first of their kind, while at the same time moving beyond that experience and on into the rest of their lives. But in their cases, the eternal question 'What next?' carried a special poignancy in a context in which the Moon programme had been shut down even before the last scheduled rockets flew.

It wasn't meant to be that way, of course. What was meant to come next had been clearly outlined. In September 1969 the NASA Space Task Group charged with directing America's space programme issued a paper, 'The Post-Apollo Space Program – Directions for the Future'. It began, 'The Space Task Group in its study of future directions in space, with recognition of the many achievements culminating in the successful flight of Apollo 11, views these achievements as only a beginning to the long-term exploration and use of space by man.' It went on to praise the astronauts who had participated in the programme so far, saying: 'We have concluded that a forward-looking space program for the future for this Nation should include continuation of manned spaceflight activity. Space will continue to provide new challenges to satisfy the innate desire of man to explore the limits of his reach.' It then proposed a base in lunar orbit by 1976, a base on the Moon by 1978, and a space station in Earth orbit with a crew of fifty by 1980. 'We conclude that NASA has the demonstrated organisational competence and technology base . . . to carry out a successful program to land man on Mars within 15 years' – in other words by 1984. By the mid-1980s, the report indicated, hundreds of people would be living in space.

Well, here we are, more than half a century later. There is still no base in lunar orbit, and no base on the Moon. The ISS, in Earth orbit, tends to host a crew of seven, not fifty, and certainly not hundreds. And for all NASA's 'demonstrated organisational competence and technology base' all those years ago, the only person to go to Mars recently was Matt Damon in *The Martian*.*

All of which dissipation turned into a lifelong predicament for the Apollo astronauts. The tide of spaceflight exploration rapidly receded around them and left them exposed, isolated for the world to marvel at, and yet, nevertheless, adrift. They found they had become the end of a story when they were hoping to be only its beginning.

'When the last of us is gone,' Gene Cernan said mournfully in 2012, 'it will all be told in the third person. It will all be hearsay.'

However, finally we can confidently state that a new era is about to dawn. The Moon is on the flight plan again, with Mars beyond it, and the stories that stopped being told for so long are, at long last, set to resume.

And soon we will have new first-person accounts to steep ourselves in, of launches and voyages and returns.

Leaving us only to wonder: what kinds of astronaut will bring them to us?

* In 2015. And he didn't seem to enjoy it very much.

EPILOGUE

In April 1960, James Van Allen, the astrophysicist who via the Explorer satellites in the early 1950s identified the radiation belts in the Earth's atmosphere – and thereby made what is regarded as the first major scientific discovery of the space age – spoke at a symposium at the Iowa Academy of Science, where he set out his firm opposition to the use of humans in space exploration.

America had recently started training its first astronauts and Van Allen felt this was plainly a mistake. He thought – and he was by no means alone in this view at the time – that there was nothing humans could do in space that couldn't be achieved more economically and with less risk by automated systems.

A solid advocate for space exploration, yet describing himself as 'a member of the loyal opposition' in this particular matter, Van Allen held on to that opinion until the end of his life, even as the tide of history set off in the opposite direction. In 2006, not long before his death, he was still arguing against the creation of the International Space Station, which was, by then, six years into its working life.

So why get into a fight with James Van Allen? Why back humans here? Humans, as we've seen in the preceding pages, are expensive and create trouble. They're messy and temperamental

301

and don't always do what you want them to do. Some of them even have (whisper it) *ideas of their own about how things should be done.*

Besides, haven't we already sent the entirely uncrewed James Webb Space Telescope a million miles out into the void? And isn't it supplying us with streams of game-changing new wisdom in the areas of cosmology and astronomy? Moreover, haven't robots already made it to Mars? Why not avoid the expenditure and the risk and all the additional complications of this proposed new twenty-first-century Artemis Moonshot programme, and beyond it the promise of onward voyages to the Red Planet, and just send out more probes?

Well, as I hope I've shown in this book, I happen to think the reasons are plentiful. And one of them, in fact, has to do with efficiency and runs quite contrary to popular perception. There's no question that robotic missions pave the way for space exploration: they have been crucial in beating the paths that humans can eventually follow. Probes reached the Moon years before Neil Armstrong and Buzz Aldrin did.

But it remains the case that what a human could do in one day on, for example, the surface of Mars in terms of data collection, surveying the landscape, picking up and analysing samples and reporting back, a robot might take as much as three years to accomplish. In space, humans frequently boast the upper hand, efficiency-wise.

And then there's our innate human desire to explore, to push the boundaries and find out what's beyond the next mountain range, across the ocean, beyond the skies above us. I recently returned from a trip to the Grand Canyon – my second visit. As I stood at the edge and peered 1,000 feet down at the Colorado river, snaking its way around Horseshoe Bend, I was struck by a sense of awe and a connection to the landscape. The sights, the smells, the sounds and that epic vista, interrupting the pace and noise of modern life and for a few precious moments allowing me to be a part of nature, the planet, the universe. I could look at a thousand photographs of the Grand Canyon, watch hours of documentary footage, but it would not remotely compare to the experience of actually standing at its very edge. And if I can't make that journey, I want to hear about it from someone who has.

In terms of almost everything we hope to get out of a spaceflight humans have shown themselves to have the edge – efficiency, connection, engagement, inspiration and, every now and again, good jokes. With all due respect to Van Allen – and hopefully the stories in this book back me up here – it's got to be humans over probes, every time.

But it takes a certain kind of human, it's true. Or rather, certain kinds of human. As I write this, we don't yet know the names of the astronauts who will crew Artemis III, the mission intended to take humans to the lunar surface for the first time since 1972. But we have been told something about its make-up. Artemis III has pledged to deliver to the Moon a woman and a person of colour.

It's a shame Edward R. Morrow didn't live to see it. In September 1961, just four months after Alan Shepard's first launch, Morrow, an American broadcaster and former war correspondent but at that point working in his role as director of President Kennedy's United States Information Agency, wrote to James Webb, NASA's first boss. 'Why don't we put the first non-white man in space?' Morrow suggested, adding, 'We could re-tell our whole space effort to the whole non-white world, which is most of it.'

The logic was emphatic, and delivered at the very dawn of the human spaceflight tale. Yet somehow America did not produce an African American astronaut for another twenty-two years, when Guion Bluford flew in the eighth Space Shuttle crew. And it was almost a decade after that, in 1992, when the first African

American woman flew – Mae Jemison, who served as a Mission Specialist on Space Shuttle *Endeavour*.

Still, the astronaut corps did eventually take steps to diversify – and in every sense. In the preceding chapters, we saw how the astronaut job description exponentially expanded as the story progressed: how the exclusive requirement for military test pilots eventually relaxed, as the scope and duration of missions changed, to admit engineers, geologists, meteorologists; how the skill-set grew in the space station and Space Shuttle age to incorporate languages, maintenance work, dentistry and doctoring; how, once the ISS was established, the ability to operate an experimental furnace in low gravity was looked for, but also the ability to fix the ISS loo when it played up and maybe do a bit of CPR if necessary, too; how astronaut became 'the ultimate portfolio role', as I described it.*

That portfolio is only going to expand again in the Artemis era. Those crews are going to need to combine the best qualities of the Apollo era with the best qualities of the long-duration ISS era. True, the job of flying the spacecraft is going to be far simpler than the last time a human crew set out for the Moon. Consider the leaps taken by computing and electronics since then. And, yes, it's an old and well-rubbed point, but it's true: your smartphone contains more computing power than an Apollo spacecraft had access to.

So tomorrow's astronauts will be devoting less of their energy to raw piloting. They're unlikely to be finding themselves, in the manner of the Apollo crews on occasions, clenched in concentration and thumbing through the manuals while unrecognised alarm bells ring. They probably won't have to spend much time,

---

* Just to be clear, CPR has not yet been necessary on the ISS, but you can trust that everybody up there knows how to administer it.

as poor, doomed Vladimir Komarov did, drumming their feet on the wall of their capsule in a vain effort to get a solar panel to deploy.

Of course, the crew will still need that essential test pilot component. These will be test flights, after all, for the Orion spacecraft and Lander Module and for hundreds of pieces of equipment as yet untried in a mission setting. The test pilot mentality will remain crucial.

But increased automation and advances in artificial intelligence in the spacecraft will free up capacity in the astronaut. And there will be a lot to fill that new capacity.

These new Moon explorers will be scientists, of course. Space exploration in the twenty-first century is, first and foremost, a scientific endeavour, and we are on the brink of an era in which space can become a large part of the solution to the problems we so clearly have on Earth. If we seize this opportunity, space can be the laboratory for many of the transformational technologies that we're going to need on our planet and a vital contributor in areas such as energy, medicine, nutrition, agriculture, robotics and clean tech. The Artemis crews will be at the forefront of that work.

They won't just be walking on the Moon, they will be constructing an Artemis Base Camp, designed eventually for, perhaps, two-month stays or longer by future crews and intended to become a vital station for the eventual push to Mars. So they're going to need to be structural engineers, experts in solar power and renewables, electrics and botany . . .

And there will be an increased call for them to be communicators and educators – something which was no real consideration for the Mercury astronauts and which has only grown to be a component of the job relatively recently, during the ISS age. Even during my selection for ESA, there was no sense that any kind of performing background might be useful to you. On the contrary, I had been flying Special Forces operations; you are very specifically discouraged from communicating about those and specifically trained in ways *not* to divulge things. But communication and outreach was very much a key part of my ISS mission – and what's the point of space exploration if it doesn't explain itself?

All in all, the job will grow again. The portfolio will expand and the challenges rise.

But on the plus side, they may, at least, have an easier time with the dust. As the Apollo crews quickly discovered, lunar regolith is not, thankfully, piled deep enough for a Lunar Lander to sink into, but it is nastily sharp and also electro-magnetically charged, for that extra cling factor, and it gets everywhere. As Gene Cernan reported, aghast: 'I have never seen so much dirt and dust in my life.' In a classic photo taken by Harrison Schmitt in the Lunar Module after a hard day's EVA, Cernan – helmets shelved beside him, his white undershirt and forehead smudged with sweat and black smears – looks less like a spaceflight pioneer and more like a coal-miner at the end of a long shift.

The good news for the Artemis crews is that NASA's Lunar Surface Innovation Consortium has been doing R&D in a long list of areas for the missions ahead, including 'active and passive mitigation strategies' for lunar dust. Apparently, pressurised liquid nitrogen has been found to get to those stubborn Moon-dust stains other detergents struggle to shift, but you probably wouldn't want to be carrying huge quantities of pressurised liquid nitrogen on a spacecraft, so I'm not sure where that leaves us. They'll think of something, though, we can be sure.

The Artemis crews are also going to have some pretty neat outfits to wear, at least based on the sightings so far of the Axiom Space Artemis III Moon Mission Spacesuit. (Needless to say, the prototype had its own streamed 'product reveal', with Peggy Whitson in attendance, plus assembled suit experts and a cast of curious schoolchildren armed with questions.) The Artemis Moonwalkers can look forward to helmet-mounted HD video cameras, far more sophisticated articulation and flexibility around the knees, elbows and finger joints compared with their Apollo predecessors – both squats and lunges will be possible, it seems – and boots designed to meet the challenge of the cold in the Moon's hitherto unexplored, permanently shadowed regions.

Their kit will also be less bulky, especially the life-support backpack, which looks like a mere school bag by comparison with the fridge-freezer-sized units of the Apollo era. That said, the sporty blue and orange of the prototype will have to become the traditional all-white during actual EVAs, the Moon being very much like the All England Tennis Club at Wimbledon in this regard (and for heat deflection purposes).

Perhaps they'll even be allowed to pack a proper golf club this time, so nobody has to do what Al Shepard did on Apollo 14 and momentarily repurpose a space tool as a six-iron. Shepard claimed forever afterwards that the two golf balls he hit that day flew for miles, the best and truest drives he ever struck – though Gene Cernan later maintained that on a photograph taken at the time he could make out two white spots only about 30 feet from the Lunar Lander, so maybe golfing yarns take on a fictional dimension even on the Moon.

The astronaut job description is set to evolve again when we look beyond the Moon to the composition of crews for Mars. What's *that* going to take? Perhaps it's best to frame this in terms of the classic film set-up: a plane comes down in a remote area and its passengers have to find a way to survive. But in this case you get to pick the dream team for the job in advance, all bases covered.

The selectors won't be building from scratch, of course. There's already more than half a century of flown experience in lunar and long-duration orbital missions which is there to be drawn on and extrapolated from, and Artemis will have added to the knowledge pile again.

At the same time, it seems clear that the Mars project is going to be placing wholly new psychological demands on its astronauts. There is going to be a vast difference between being thrown into low orbit on the ISS for a year and being able to go to the window and see a giant Earth turning underneath you, and sitting in a spacecraft and watching that same comforting Earth diminish until it becomes a pinprick of light in the blackness.

Similarly, you're going to be living for maybe three years with the knowledge that there is no easy way home. Even on the ISS, you could always tell yourself that in the event of the direst of outcomes (a fire, a major depressurisation) there was a lifeboat – that Soyuz capsule, docked outside. From Mars there really will be no quick way home in an emergency. The Mars crews will be living with a sense of isolation and enclosure at an altogether new and uncharted level.

On the ISS, you are kept pretty busy, and surrounded by equally busy colleagues. You're not given much slack time in which you might settle into yourself. Will they be busy during the eight or nine months of the long push to Mars? Or will boredom be a factor?

And then what about the communications delay? The Mars crew can expect anything up to a twenty-minute lag between their messages to and from Earth. During the twelve-day underwater NEEMO expedition that I undertook as part of my astronaut training a ninety-second lag was artificially introduced between us and our mission controllers at the surface. Even that relatively short space between question and answer introduced levels of chaos and proved weirdly spirit-sapping. Twenty-minute gaps are going to entirely remodel the nature of the collaboration between crew and base, creating scenarios in which the crew learns to unplug and work autonomously, without the constant input of Mission Control.

How do you prepare for all this? Simulations may help. NASA has, at the time of writing, begun a long-term isolation experiment, and ESA has already conducted the Mars-500 experiment in long-duration mission endurance. On 3 June 2010 six participants (three Russian, one French, one Chinese and one Italian) entered a 72-square-metre spaceship mock-up in Russia's Institute for Biomedical Problems in Moscow, and then had the hatch bolted behind them. They spent eight months 'journeying' to 'Mars', they ate space food, they conducted scientific experiments in the spacecraft's lab. In February 2011, three of them boarded a Lander and ventured to the Martian 'surface' for ten days, conducting a number of EVAs on what was actually a 10-foot by 6-foot stage-set. Then they all reunited and 'journeyed' back again, emerging 520 days later on 4 November 2011 in a snowy Moscow. All the while medics had been monitoring their immune systems, sleep cycles, hormone levels and moods, and harvesting all the data they could.

At the same time, the fundamental artificiality of those scenarios is always going to come into play at some point. As an Army officer, I underwent a 'Conduct after Capture and Resistance to Interrogation' training exercise, including strip search, dousings with freezing cold water, cell confinement, stress positions – the full enchilada. I've had nicer evenings, no question. I've had nicer mornings, afternoons and nights, too. (It went on for forty-eight hours.) And I'm sure the experience was a useful eye-opener in terms of what I might have expected to go through in the kind of situation which I was fortunate never to get into.

And it would have been even more of an eye-opener, I guess, if I hadn't spent so much of it with a sack over my head.

At the same time, though, even in those hyper-realistic role-playing scenarios, you always have access to a little voice inside you which says: 'It's OK – it's only pretend. They're not actually going to kill you, even though they keep telling you that they are. When this is over, you're going to be able to drive to McDonald's on the way home and order a quarter-pounder with cheese, some fries and a milkshake.' (Which, as it happens, is exactly what I did.)

My point is, there are certain experiences it's impossible completely to prepare yourself for, and heading off to Mars is certainly going to be one of them.

But boldness in the face of the unknown – and, indeed, *excitement* in the face of the unknown – is something which astronauts have been carrying with them to the launch pad since spaceflight began. And two things can be said about that Mars mission for sure.

Firstly, that it will be the most ambitious journey of exploration ever undertaken and that, as such, it will be arduous, complex, fraught, uncertain and at times possibly even quite terrifying for the humans involved.

And secondly, that every person who has ever been an astronaut will be wishing they were going too.

ACKNOWLEDGEMENTS

This book would not have been possible without the invaluable conversations and insights I have gathered from my fellow astronauts, colleagues, friends and experts who I have had the privilege to collaborate with and learn from during my career as an astronaut. There are too many names to mention, but I would like to thank everyone at ESA, NASA, the UK Space Agency, JAXA, CSA and Roscosmos, who have collectively shaped my view of the human story of space exploration. I would like to thank Giles Smith, Nick Spall and everyone at my publisher Penguin Random House for their hard work in bringing this book into the world: Ben Brusey, Joanna Taylor, Jason Smith, Jessica Fletcher, Anna Cowling, Sarah Ridley, Klara Zak, Tory Lyne-Pirkis and Francisca Monteiro. I would like to thank my literary agent Julian Alexander and everyone at The Soho Agency for all of their support. Finally, thanks to my wife, Rebecca, and my sons, Thomas and Oliver, for inspiring me and for being there for me every step of the way, as always.

BIBLIOGRAPHY

Anyone who writes about spaceflight, like anyone who becomes an astronaut, is aware of walking in the footprints of many exceptional forerunners. All of the books in the following list have informed my sense of spaceflight's history and influenced my thinking about it, as well as entertained me royally, and I want to acknowledge my debt and gratitude to all of their authors. I also strongly recommend every one of these volumes to anyone keen to take their interest in this subject further.

I would also like to express my gratitude to the keepers of NASA's simply magnificent online archive, and to the editors and writers of *National Geographic*, the *Smithsonian* magazine, the *New Yorker* and *New Scientist*.

Aldrin, Buzz, *Mission to Mars*, National Geographic Society (2013)

Aldrin, Buzz, with Abraham, Ken, *Magnificent Desolation: The Long Journey Home from the Moon*, Bloomsbury (2009)

Ansari, Anousheh, with Hickam, Homer, *My Dream of Stars*, St Martin's Press (2010)

Bean, Alan, *Apollo: An Eyewitness Account*, Greenwich Workshop Press Inc. (2002)

Burrough, Bryan, *Dragonfly: NASA and the Crisis Aboard Mir*, HarperCollins (1998)

Cadbury, Deborah, *Space Race*, Fourth Estate (2005)

Cernan, Eugene, with Davis, Don, *The Last Man on the Moon*, St Martin's Press (1999)

Chaikin, Andrew, *A Man on the Moon*, Penguin (2019)

Collins, Michael, *Carrying the Fire*, Farrar, Straus & Giroux (1974)

Conrad, Nancy, with Klausen, Howard, *Rocketman: Astronaut Pete Conrad's Incredible Ride to the Moon and Beyond*, New American Library (2005)

Dean, Margaret Lazarus, *Leaving Orbit: Notes from the Last Days of American Spaceflight*, Graywolf Press (2015)

DeGroot, Gerard, *Dark Side of the Moon: The Magnificent Madness of the American Lunar Quest*, Jonathan Cape (2007)

Doran, Jamie, and Bizony, Piers, *Starman: The Truth Behind the Legend of Yuri Gagarin: 50th Anniversary Edition*, Bloomsbury Paperbacks (2011)

Dubbs, Chris, and Paat-Dahlstrom, Emmeline, *Realizing Tomorrow: The Path to Private Spaceflight*, University of Nebraska (2011)

Foale, Colin, *Waystation to the Stars: The Story of Mir, Michael and Me*, Headline (1999)

Foster, Amy, *Integrating Women into the Astronaut Corps – Politics and Logistics at NASA, 1972–2004*, Johns Hopkins University Press (2011)

French, Francis, and Burgess, Colin, *In the Shadow of the Moon: A Challenging Journey to Tranquility, 1965–1969*, Bison Books (2010)

French, Francis, and Burgess, Colin, *Into That Silent Sea: Trailblazers of the Space Era, 1961–1965*, Bison Books (2010)

Gagarin, Yuri (as told to N. Denisov and S. Borzenko, edited by N. Kamanin, translated from the Russian by G. Hanna and D. Myshne), *Road to the Stars*, Foreign Languages Publishing House (1962)

Hadfield, Chris, *An Astronaut's Guide to Life on Earth*, Little, Brown (2013)

Hall, Rex, Shayler, David, and Vis, Bert, *Russia's Cosmonauts: Inside the Yuri Gagarin Training Center*, Springer Praxis Books (2005)

Hansen, James R., *First Man: The Life of Neil A. Armstrong*, Simon & Schuster (2018)

Hitt, David, Garriott, Owen, and Kerwin, Joe, *Homesteading Space: the Skylab Story*, Bison Books (2011)

Jackson, Libby, *A Galaxy of Her Own: Amazing Stories of Women in Space*, Cornerstone (2017)

Kluger, Jeffrey, *Apollo 8: The Thrilling Story of the First Mission to the Moon*, Henry Holt (2017)

Koppel, Lily, *The Astronaut Wives Club*, Headline (2013)

Kraft, Chris, *Flight: My Life in Mission Control*, Dutton (2001)

Kranz, Gene, *Failure Is Not an Option*, Simon & Schuster (2000)

Leinbach, Michael D., and Ward, Jonathan H., *Bringing Columbia Home*, Arcade (2018)

Light, Michael, *Full Moon*, Jonathan Cape (2002)

Lovell, James, and Kluger, Jeffrey, *Lost Moon: The Perilous Voyage of Apollo 13*, Houghton Mifflin (1994)

Mailer, Norman, *Of a Fire on the Moon*, Little, Brown (1969)

Mercury Seven, *Into Orbit*, Cassell & Co. (1962)

Mullane, Mike, *Riding Rockets: The Outrageous Tales of a Space Shuttle Astronaut*, Scribner (2007)

Potter, Christopher, *The Earth Gazers*, Head of Zeus (2017)

Riley, Christopher, and Impey, Martin, *Where Once We Stood: Stories of the Apollo Astronauts Who Walked on the Moon*, Harbour Moon Publishing (2019)

Saunders, Andy, *Apollo Remastered*, Particular Books (2022)

Sharman, Helen, with Priest, Christopher, *Seize the Moment*, Victor Gollancz (1993)

Sherr, Lynn, *Sally Ride, America's First Woman in Space*, Simon & Schuster (2014)

Slayton, Donald K., with Cassutt, Michael, *Deke! From Mercury to the Shuttle*, Forge (1995)

Smith, Andrew, *Moondust: In Search of the Men who Fell to Earth*, Bloomsbury (2019)

Turnill, Reginald, *The Moonlandings: An Eyewitness Account*, Cambridge University Press (2002)

Walker, Stephen, *Beyond: The Astonishing Story of the First Human to Leave Our Planet and Journey into Space*, William Collins (2021)

Wolfe, Tom, *The Right Stuff*, Vintage (1979)

Zimmerman, Robert, *Genesis: The Story of Apollo 8*, Dell (2000)

INDEX

INDEX

INDEX

PHOTOGRAPHY CREDITS

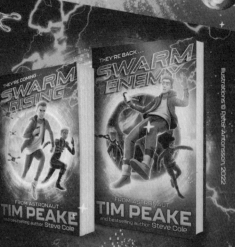